SUSTAINABILITY AND DEGRADATION IN LESS DEVELOPED COUNTRIES

*For my daughters Serena and Clea,
and for all the children everywhere,
now and in the future.*

Sustainability and Degradation in Less Developed Countries

Immolating the future?

SARAH LUMLEY
The University of Western Australia

Routledge
Taylor & Francis Group

LONDON AND NEW YORK

First published 2002 by Ashgate Publishing

Reissued 2018 by Routledge
2 Park Square, Milton Park, Abingdon, Oxon OX14 4RN
711 Third Avenue, New York, NY 10017, USA

Routledge is an imprint of the Taylor & Francis Group, an informa business

Copyright © Sarah Lumley 2002

The author has asserted her moral right under the Copyright, Designs and Patents Act, 1988, to be identified as the author of this work.

Publisher's Note
The publisher has gone to great lengths to ensure the quality of this reprint but points out that some imperfections in the original copies may be apparent.

Disclaimer
The publisher has made every effort to trace copyright holders and welcomes correspondence from those they have been unable to contact.

A Library of Congress record exists under LC control number: 2002018634

ISBN 13: 978-1-138-72845-5 (hbk)
ISBN 13: 978-1-315-18998-7 (ebk)

Contents

v

List of Figures and Plates

Figures

Plates

List of Tables

Hypothesis Tables

Preface

As the title suggests, this is a book about sustainability. While the main focus of the work is on less developed countries generally, and on the Philippines in particular, it has a strong global perspective. This perspective is in recognition of the fact that nothing on earth happens in isolation, and that the evolution of theories and concepts in sustainability, covered briefly in this book from their conception to the present, have relevance to all nations. It also acknowledges that understanding the experience of one nation has relevance to another.

The empirical work presented here relates to a case-study of farmers facing land degradation problems at four upland sites on the Philippines' island of Leyte. In a broad sense the analysis examines the socio-economic factors that influence farmers' decision-making about soil conservation adoption, with particular attention to farmers' private time preference rates, discounting theory and the economic assumptions that underpin such theory.

Biogeophysical attributes and the history of the Philippines are also considered, in recognition of geographical, historical, cultural and political influences on decision-making, and on social and environmental sustainability. Perceived flaws in economic theory, and in its assumptions, interpretations and applications, are discussed in the context of promulgating socially and environmentally sustainable policies for the Philippines and beyond. Controversies embedded in global sustainability dialogues and debates are recognized and briefly discussed.

It is hoped that this book can make a small contribution to the development of theories and policies that can help to ensure, for the majority of the world's people, a future that is better than the present.

Acknowledgements

Numerous people have made direct or indirect contributions to this book, including my teachers, my students, my colleagues, my friends and my family. They are too many to mention by name but I would like my thanks to go to them all. Particular acknowledgement is due to the following: Dr Bill Stent, Director of Economics in the former Victorian Department of Conservation and Environment; my friends and former colleagues there, Annette Bould and Athena Andriotis; Professor Greg O'Brien, Dean of Law and Management at La Trobe University; Dr Rob Dumsday, La Trobe University; Ms Claire Thomas, Victorian Department of the Treasury; Mr Mike Smith, Former Director of Resources at the Department of the Premier and Cabinet; former colleagues at the EPA, Victoria; colleagues at ViSCA, Leyte; former colleagues in Geography at Deakin University; past and present colleagues in Geography at the University of Western Australia (UWA) including Associate Professor Dennis Rumley; Dr Brian Shaw; Dr Ian Alexander; Dr Marion Hercock; my colleagues in the School of Agricultural and Resource Economics at UWA; friends and colleagues in other UWA departments including Anthropology, English, Philosophy, Botany, Soil Science, and Asian Studies; research colleagues Associate Professor Pierre Horwitz and Dr Sue Nikoletti at Edith Cowan University, and Dr Matthew Tonts, UWA; Mr Jonathan Thomas, The Resource Economics Unit, Subiaco; Dr Jay Gomboso, WA Department of Conservation and Land Management; Professor Rob Fraser, London University; Dr Laura McCann, Columbia University, Missouri; my PhD students, Annette Baumann, Trudy Hoad, Fiona McKenzie, Ron Sheen and Nuri Dewi Yanti; Masters student Allan Dobra; Ian Baker; Lal and Cedric Baker; Neil and Molly Garland; Joan Phillips; Dr Charlotte Brack; Doris Tate and Graeme Henry; Bette Moore; Mark Ducksbury; Leni Brockmuller; Barbara Pedersen; Andrea Hinwood; Cathy and John Loughridge, and Dr Margaret Lumley. Special thanks are due to Andrew Kemp for rescuing my graphics. Lastly I must recognise the support of my sisters Philippa Thornton and Caroline Lumley; my parents Roma and Reg Lumley; my nieces Kate, Juliet, Emma and Georgina; my nephews James and Jonathan; and, of course, my delightful and long-suffering daughters Serena and Clea, who are always supportive of their mother's desire to write arcane texts.

List of Abbreviations and Acronyms

ACIAR	Australian Centre for International Agricultural Research
CBA	Cost-benefit analysis
CV	Compensating variation
EV	Equivalent variation
IRR	Internal rate of return
MOC	Marginal opportunity cost
NPA	New People's Army
NPV	Net present value
PTPR	Private time preference rate
STPR	Social time preference rate
TPR	Time preference rate
UNCED	United Nations Conference on Environment and Development
ViSCA	Visayas' State College of Agriculture
WCED	World Commission on Environment and Development
WTA	Willingness to accept (compensation)
WTP	Willingness to pay

'... where natural resources are concerned, we sacrifice a pretty accurately predictable future to present greed. We know, for example, that if we abuse the soil, it will lose its fertility, that if we massacre the forests, our children will lack timber and see their uplands eroded, their valleys swept by floods. Nevertheless, we continue to abuse the soil and massacre the forests. In a word, we immolate the present to the future in those complex human affairs, where foresight is impossible; but in the relatively simple affairs of nature, where we know quite well what is likely to happen, we immolate the future to the present.'

ALDOUS HUXLEY
Time Must Have a Stop (1944, p. 266)

Chapter 1

Sustainable Development and Land Degradation

Sustainable Resource Use

The work presented in this book makes a broad, multidisciplinary exploration of a range of issues that contribute to the unsustainable use of natural resources. This work has a particular focus on sloping agricultural land degradation in the Philippines. It considers the effects of hill farmers' socio-economic circumstances on their conservation adopting behaviour and on their decision-making. The book is set in the overall context of sustainable development. This introductory chapter begins with a brief synopsis of the evolving theory of sustainable development and its application to social and environmental policy.

It is well recognised that a 'green' discourse which embodied aspects of sustainable development, including the still controversial 'limits to growth' arguments, existed in the 1970s (Buttel, 1998; Howes, 2000). Yet even further back, and following a similar rationale, some political economists, philosophers and scientists were supporting parallel notions of moral duty and applied utilitarianism with respect both to society and the environment.

For example, Harriet Martineau, a 19th-century political economist and journalist, translated via the popular press the work of intellectuals like Adam Smith, Thomas Malthus and David Ricardo for the general population. Martineau believed in and espoused concepts of political economy and moral duty that bear a striking resemblance to today's ideas about social and environmental sustainability. Yet, even in the 19th-century political economy as a means of determining policy was not without controversy. The poet William Wordsworth and the philosopher Thomas Carlyle, both friends of Martineau, expressed great concern that political economy would turn moral values into a 'cost calculus', and Smith, Malthus and John Stuart Mill were all '... at great pains to conserve "moral

1

sentiments" and higher human values in the context of their [own] narrowly based proposals for utilitarian economics' (Lumley, 2000 (a), p. 63).

Notions of sustainability were promoted at the United Nations Conference on the Human Environment in Stockholm in 1972. Following the Stockholm Conference, a series of policy agreements was made. These agreements covered issues as diverse as ocean pollution, air pollution and endangered species (Elliott, 1998; Buttel, 1998; Howes, 2000). However, the term sustainable development only became popular in 1987 after the World Commission on Environment and Development published *Our Common Future* (WCED, 1987). Also known as the Brundtland Report after the WCED chairperson, this publication made, for the first time, a publicly heard call for a global strategy which linked economics with the environment. Although some of its terms and definitions were confusing and ill-explained, for example the words 'growth' and 'development' were used interchangeably, this report generated great interest in the topic of environmental sustainability. *Our Common Future* supported 'sustainable development' as a means of resolving poverty, pollution, inequality, loss of biodiversity, land degradation and the depletion of natural resources, thus juxtaposing social and economic well-being with environmental well-being. The report (*ibid*, p. 43), in a well meant phrase that has become something of a cliché, defined sustainable development as: 'development that meets the needs of the present without compromising the ability of future generations to meet their own needs'.

Our Common Future was followed by another publication, *Blueprint for a Green Economy* (Pearce *et al*, 1989), which also caught the imagination of economists, conservationists and the general public, and which made a definitive attempt to link the economies and the environments of nations globally. In *Blueprint*, proposals were made in which deals between developed and undeveloped countries, such as debt-for-nature swaps, could be worked co-operatively to resolve economic and environmental problems. Many of these ideas were neither completely new nor entirely practical, but their consolidation in a popular book helped to bring sustainable development discussions to a wider forum than hitherto had been the case.

By the late 1980s there was much literary activity in disciplinary areas relevant to most aspects of sustainability. Following the release of the Brundtland report, a new journal, and what some considered to be a new discipline, *Ecological Economics*, emerged. Here, among other issues, economists and ecologists sought to develop ways in which solutions to the problem of environmental degradation could be found. Many journal articles and books which addressed topics that coupled economics and sustainable environmental use were published in the five years that followed the first release of the Brundtland Report (e.g.Tietenberg,1988; Pearce, 1989; D'Arge, 1990; Turner, 1990; Hueting, 1990; Pearce and Turner, 1990; Archibugi and Nijkamp, 1990; Friend and Rapport, 1991; Pearce, Barbier and Markandya, 1990; Tisdell, 1991; and Dragun and Jakobsson, 1992).

One of the problems that these and subsequent writers have tried to resolve is how to define properly the term 'sustainable development'. This term has often been interpreted according to the interests and ideology of the interpreter, and in this respect has given little common ground for understanding and resolving the central issues involved. This has particularly been the case for those issues that concern effective social and environmental policy formulation (for example see Daly, 1990; Jacobs, 1993; Common, 1995; McManus, 1996; Bowers, 1997; Lumley, 1999).

In Australia, as in many other countries, there has been much concern about soil erosion and salinity, pollution of air and water resources, and what some see as unsustainable use of forest resources. In 1990, with the publication of its report on sustainable development, the then Australian Government initiated a process which aimed to develop and implement a strategy for 'ecologically sustainable development' by 1992 (Commonwealth of Australia, 1990).

In 1992 WCED supported the now renowned United Nations Conference on Environment and Development (UNCED) 'Earth Summit' in Rio de Janeiro. At Rio, the non-binding agreement, Agenda 21, a global agenda for the 21st century, was signed as a plan of action to implement the similarly non-binding Rio Declaration. The Rio Declaration had outlined 27 guiding principles of sustainable development for governments globally (Elliott, 1998). The United Nations Convention Framework on Climate Change, and the Convention on Biological Diversity were also non-binding UNCED documents. Despite the much publicised proposal to ratify the binding Kyoto Protocol on Climate Change in 1998, there is yet to be an international agreement which legally upholds any of these recent global sustainable development policies. However, there is now no shortage of definitions and examples of proposed actions for sustainable develpoment.

Australia, like other developed and less-developed countries, continued to pursue its own policies for environmental and social sustainability. An interim Australian strategy on sustainable development, which was published in late 1992, described (ecologically) sustainable development thus:

> In a general sense, ecologically sustainable development is about the way we use, conserve and enhance the community's resources so that

ecological processes, on which life depends, are maintained, and the total quality of life, now and in the future, is secured (Commonwealth of Australia, 1992; p. 3).

This definition was seen to reflect ... 'recognition that in Australia, economic development and a well managed environment are inextricably linked (*ibid*, p. 4). By 1993 the Australian Government had proposed a more specific definition for ecologically sustainable development: 'development that improves the total quality of life, both now and in the future, in a way that maintains the ecological processes on which life depends' (Commonwealth of Australia, 1994; p. 2).

The definition illustrates the difficulty that governments have in proposing ideas that are supposed to have some practical application. One of the problems faced by governments when trying to define such terms is that they attempt to avoid alienating interest groups, which may be opposed to each other (e.g. business and conservation groups) and so end up with a term that is difficult to apply to practical policy.

Despite continuing popularisation of the notion that economics and environmental management are linked, as mentioned earlier, this linkage has been recognised at some level for many years. Indeed there have been academic and practical arguments running on this theme as far back as the mid 19th century. Mayumi (1991) when discussing the cause of land degradation quotes a 19th-century agronomist (Liebig, 1859), who was concerned about land degradation and the exhaustion of the soil, as saying:

But who could have thought twenty years ago, when there was plenty of manure, that it would ever occur to these obstinate and wilful fodder plants to produce no more manure, and no longer spare and enrich the ground. The soil is naturally not the cause of this; for they teach that it is inexhaustible, and those still enough believe that the source from which it is derived will always flow (pp. 130-131).

Magumi argues that a thermodynamic analysis is appropriate for considering the process of soil degradation since: 'the tremendous speed of matter and energy degradation of the earth is the genuine cause of land degradation' (p. 35) and that the 19th-century agricultural economists of Europe '... did not pay sufficient attention to the importance of circulation of matter in order to maintain land fertility in the long run. Their interest was to increase the amount of crop yields in a short span of time' (p. 38). This argument has some merit.

Since *Our Common Future* was published, the discourse on sustainable development has been growing more lively, more controversial and more complex. One environmental economist who has participated fully, and over the long-term, in the environmental economics/sustainable development debate is David Pearce, co-author of the definitive *Blueprint* series. In a recent book (Pearce, 1999) David Pearce presents a collection of his essays that spans over twenty years of his work in the environmental management and policy arena. He comments:

> Few concepts have attracted so much political, popular and academic attention as that of 'sustainable development'. While politicians are adept at embracing high-sounding objectives – especially when they are so loosely defined as to be consistent with almost any form of action (or inaction) – it is significant that 'sustainable development' now figures as a goal in dozens of national environmental policy statements and in the opening paragraphs of 'Agenda 21', the massive shopping list of world actions adopted at the Earth Summit in Rio de Janeiro in June 1992.

And,

> While it is a popular pastime to collect different different and incompatible definitions of sustainable development, inspection of the words and of their origins suggests that *defining* sustainable development is really not a difficult issue. The difficult issue is in determining what has to be done to achieve sustainable development, assuming it is a desirable goal (p. 69).

Pearce then clarifies the issue for us by explaining that what we're really talking about is sustainable *economic* development [my emphasis] and proceeds with his exposition from there. Given that there is still some dispute about whether economic development is what sustainable development alludes to, and given the heat with which the word 'growth' was contested early in the debate, when it was sometimes used interchangeably with the word 'development' (Daly, 1990), it may surprise some readers to know that the matter has been resolved. Perhaps central to this issue is the question of what constitutes 'economics', and how it is applied to environmental and social policy. A common misconception is the idea that all economics is about money and profit. Another is the confusion of rationality in economic theory with the expedient of 'economic' rationalism. According to Pearce (1999, p. 14):

> These rather basic misconceptions would not matter very much but for the fact that critiques of economics and environmental economics are built up on the basis of this straw man. One suspects that the age-old confusion

between economics as commerce and, economics as science, has not yet gone away.

The assumptions that underlie standard neo-classical economic theory are often little known, little understood, and difficult to challenge. If we are to accept that economics is intended to encompass the social dimension then it must also be accepted that economics is far broader than issues concerning finance and commerce. As such it is important to social and environmental policy. This is particularly so where people are directly dependent upon the environment for their subsistence, as they are in less developed nations like the Philippines (Horne, 1996; Sponsel *et al*, 1996; Japan Environmental Council, 2000). Yet in more developed places like Australia, the European Community, Japan and North America we have come to realise that we also depend, directly or indirectly, on the environment for our survival (Gruen and Jamieson, 1994; Lumley *et al*, 2001; Ullsten and Rapport, 2001). Recent research has confimed that, in Western Australia at least, ordinary members of society perceive this dependence to be real. In a survey there, 98.5% of respondents believed that 'Looking after the environment is important to our long-term survival' and 91% believe that 'The well-being of society, the economy and the natural environment are strongly linked' (Lumley and Hercock, 2001). In this regard, environmental economics, ecological economics and the concept of sustainable development are all socially relevant. As Mishan (1979, p. 389) states of welfare economics:

> ... inasmuch as current allocation economics derives its rationale from welfare economics, the socially relevant part of that subject can have no affinity with the species of sophisticated games which economists can play with any ranking device that catches their fancy. As the term suggests, welfare economics is to be regarded as a study of the contribution economics can make to advancing social welfare.

The research upon which this book is based examines the socio-economic conditions of Philippines upland farmers in the context of their relationship to their environment, and particularly to the land that they cultivate. The work assesses the validity some of the assumptions in economic theory, and the effect of invoking such assumptions on the advancement of social welfare. In addition, the interpretation and application of some aspects of [Western] economic theory in a less developed nation are examined. This is done against a backdrop of land degradation and poverty, and uses an approach that considers the applicability of sustainability as a concept that can be used to advise environmental and social policy development.

The Problem of Land Degradation

One of the main reasons for land degradation is the use of non-sustainable agricultural practices. In many cases unsustainable agriculture begins with the removal of perennial vegetation and its replacement with shallow rooted annual crops. The cultivation of such crops often leaves the soil bare for extended periods, and results in the inexorable decline of the land upon which people depend for both their sustenance and their livelihood.

Pimental *et al* (1994) illustrate the importance of land degradation as an environmental and social problem in the following statement:

> Rapid land degradation is a major threat to the sustainability of world food supply and affects most of the crop and pasture land throughout the world ... Estimates suggest that agricultural land degradation can be expected to depress food production between 15 % and 30 % over the next 25-year period, unless sound conservation practices are instituted now ...
>
> Soil erosion is the single most serious cause of this degradation ... The major cause is the employment of poor agricultural practices that leave the soil without vegetative cover to protect against water and wind erosion. Soil loss is particularly distressing because it takes approximately 500 years to reform 2.5 cm (1 inch) of topsoil under normal agricultural conditions (p. 203).

As with most environmental problems, the recognition of land degradation as a pressing issue in human ecology has long been recognised. In E. Ray Lankester's popular essay 'The Effacement of Nature by Man' published in 1913 and reissued recently (Lankester, 2000), the author examined the implications of human induced environmental damage, and particularly desertification, over the entire globe, but with a focus on its implications for less-developed countries. He commented:

> Sand-deserts are ... areas of destruction of vegetation – often (though not always), both in Central Asia and in North Africa (Egypt etc.), started by the deliberate destruction of forest by man, who has either by artificial drainage starved the forest, or by the simple use of axe and fire cleared it away (*ibid*, p. 238).

There has been an increasing concentration on the theory and practice of vegetation retention or replacement, and on resource conservation generally, in less-developed countries. The focus of such work has been on the amelioration of a number of environmental problems including land degradation, and to recognise broader ecosystem values (for example Palo and Mery, 1996; Perna and Santos, 1998; Lumley, 1998 (a); Chivaura *et al*, 2000; Godoy *et al*, 2001; Guo *et al*, 2001). Yet developed countries are also sometimes struggling with land degradation in forms such as salinity, erosion and desertification, even while devegetation and commercial logging activities vie with sustainable development strategies for a place in state and national policy annals (for example, Mueller *et al*, 1999; State Salinity Council, 2000; Kennedy *et al*, 2001; Grigalunas *et al*, 2001). The reasons for devegetation, land degradation, the over-exploitation of natural resources, and the frequent inability of farmer decision makers to adopt agricultural practices known to be sustainable may be similar in both developed and less-developed nations.

In the Philippines there was a 2 % annual decline in forest cover from 1981 to 1990 (Kidron and Segal, 1995), a rate which has probably accelerated in the subsequent decade. And in Western Australia a political battle erupted, and continues to flare intermittently, over a State/ Commonwealth Government bilateral agreement to log remnant old-growth forest in a part of the country that is estimated to have already lost about 6 million hectares of land to dryland salinity (JCWRFAFC, 1998; Pannell, 2001).

Soil and Slope Erosion in the Philippines

The problem of soil erosion occurs all over the Philippines and results in physical and financial hardship in the short term. In the long term its consequences may be even more far-reaching.

Many of the slopes of the Philippines have been cleared of forests whose mineral and organic matter, absorbed from the soil over a long period, was not returned to it at the time of deforestation, so that the soil on the newly cleared agricultural land was already destabilised when ready for cultivation. As Cramb and Saguiguit (2000) state:

> Human-induced soil erosion and consequent degradation of agricultural land in the Philippine uplands has, during the past three decades, come to be recognised as a major environmental and socioeconomic problem (p. 9).

By 1981 it was estimated that 9 million hectares of alienable and disposable land in the Philippines (about 69% of such lands) were eroded (Cabrido, 1981). Sajise and Ganapin (1990) say of the Philippines:

> ... there is a general recognition of the serious problems associated with deforestation, soil erosion, declining agricultural productivity, loss of

biodiversity, off-site impacts, the increasing poverty and the social costs associated with the bio-physical and ecological instability of the uplands (p. 31).

The upland areas of the Philippines cover about 50% of the total land area. Nearly all of the Philippines' remaining forests are in the uplands, but in the past half century agriculture has steadily encroached on the steep forest lands. In Zamboanga del Sur forest clearing started after World War II and now only 16% of the original forest cover remains. In the 15 years to 1985 farmers claimed that there was an 80% decline in corn yield and up to one metre of soil depth was lost (O'Sullivan, 1985).

In the Philippines nearly 19 million people live in the uplands, with 8.5 million living in the forests (Umali, 1991). One characteristic of the uplands is that they cannot sustain intensive agriculture. According to Watson (1983):

> There is an urgency to the problem of land degradation. It has been stated that two thirds of the world's population is living in developing nations. Agricultural production is not keeping pace with population increase. It has been estimated that about three – fourths of the farmers in the developing nations are hillside farmers. Hillside lands are eroding at a staggering rate. This erosion problem is causing untold hunger, poverty and damage to the people of Asia and Southeast Asia and the Philippines in particular. The economic survival of developing nations may depend upon what is accomplished in the next ten years in sloping land degradation (p. 112).

This statement was made nearly twenty years ago and there is no evidence of improvement. Indeed the problem may have accelerated. It is apparent that the Philippines is unable to match increased crop production with population growth, experiencing a drop from 91.2 % to 86.9 % in grain self-sufficiency between 1980 and 1990 (Japan Environmental Council, 2000).

Deforestation is the main cause of soil erosion everywhere in the world, and tropical deforestation is contibuting to degradation, loss of autonomy and human suffering in Central and South America, Africa, India, and New Guinea and the wider Pacific, as well as in the Philippines (Sponsel *et al*, 1996). Inappropriate land use and bad soil and crop management have also contributed to the problem. The soil lost from the degrading uplands in the Philippines leads to siltation of waterways and flooding in the valleys below (Santa Maria Maniego, 1986; Horne, 1996; Cramb, 2000).

The Mechanics of Slope Erosion in the Philippines

The reason for slope erosion in the Philippines is the exposure of unstable soils during periods of heavy rainfall. Usually, the longer the duration of intensive rainfall the more extensive the soil damage (Rao, 1972). Lack of vegetative cover means that the exposed soil is undermined by the rain and carried away by runoff of water which would normally be absorbed by vegetation. Generally, the more runoff that occurs, the greater the extent of soil loss. Soil erosion also destabilises the slopes so that landslides can occur. Mud flows that result from such landslides frequently cause serious damage (Watson, 1983). According to Santa Maria Maniego (1986):

> ... the amount of soil transported was directly proportional to the intensity of the precipitation and showed a logarithmic relationship with drop diameter. The amount of fine sand transported by drop impact is directly proportional to the total mass of water supplied and to a factor representing the energy per unit area supplied by the individual drop (p. 4).

The nature of vegetation cover, whether it be annual or perennial, broad or narrow leafed, shallow or deep rooted, influences the impact of rainfall, runoff and soil loss. Perennial, broad leafed, deep rooted plants afford the most protection to the soil (Corpuz, 1979; Cramb, 2000). The nature of the slope also affects soil erosion. The steeper the slope the greater the velocity of water runoff. Generally, if the degree of the slope is doubled, the amount of erosion is doubled. Soil erosion is also greater on long slopes as compared with short slopes because there is a greater area for uptake of the soil by the runoff (Farris, 1948).

Finding Solutions to the Problem

Research Projects

There have been various research projects conducted in the Philippines which have investigated agricultural resources and have attempted to propose technologies suitable for improving farm efficiency and viability. An extensive project conducted in Zamboanga del Sur in Mindanao (ZDSDP, 1983) investigated living standards, infrastructure development, agricultural research, production and extension, and social development.

Because the focus of the project was so broad there was little concentration on soil erosion although it was recognised that soil loss was a problem. The solution proposed was to encourage cultivation of tree legume hedges on the steep denuded slopes (Brookes, 1983). Cultivation of tree legumes such as ipil-ipil (*Leucaena leucocephala*) appears to have become a standard proposal for slope erosion control in the Philippines (e.g. Celestino, 1985; Pacardo, 1982; Watson, 1983; Santa Maria Maniego, 1986). Research and extension projects have been conducted by Philippines government agencies such as the Bureau of Soils, the National Environmental Council and the Bureau of Forest Development who have a responsibility for soil conservation and erosion control.

Other agencies such as the Human Settlements Regulatory Commission, National Irrigation Administration, Bureau of Lands, Bureau of Agricultural Extension, Ministry of Agrarian Reform, and the Philippines Council for Agriculture and Resources Research and Development have played supportive roles over the years (Cabrido, 1985; Concepcion, 1983; Librero, 1983). Overseas aid bureaux have also become involved in assisting with uplands soil erosion mitigation in southeast Asia. For example, AIDAB (the Australian International Development Assistance Bureau), ACIAR (the Australian Centre for International Agricultural Research) and the Queensland Department of Primary Industry have displayed a particular interest in upland erosion (for example see Cramb, 2000; Rady, 1990 and Loch, 1985). North American agencies have also been involved, with the Ford Foundation and the US Agency for International Development sponsoring projects in Indonesia and the Philippines (see KEPAS, 1985 and Parrilla *et al*, 1988). Some of these projects have been conducted over several years, with the goal of introducing erosion mitigation techniques to landholders. The technologies that have been most prominent in recent research are those involving contour hedgerows and/or bench terraces (Lumley and Stent, 1989; Lumley, 1995; Cramb, 2000). Yet although it has become clear that farmers have become aware of such technologies, adoption rates are sometimes slow (Parrilla *et al*, 1988).

Dalton (1986) stated that the agricultural development process was a four stage cycle which '... should be managed by institutions sensitive to rural communities and with a maximum of farmer participation ...' and that the productive processes '... fall into four main categories i.e. annual crop, perennial crop, livestock and cottage industries and can be related in terms of the efficiency with which they achieve the four purposes of the farming system (i.e. secure staple crop yields, control erosion, diversify production and maintain fertility)' (pp. 1, 2). Despite research and extension efforts these four purposes have not been achieved in the Philippine uplands and one of the questions asked in this book is why?

Possible Solutions to Soil Erosion

One of the technologies which has been introduced to hill farmers experiencing soil erosion in the Philippines is known as SALT (Sloping Agricultural Land Technology). This was developed on the island of Mindanao, by the North American Baptist Rural Life Center (and World Neighbours) which opened in 1971, to introduce a system of upland farming for farmers with small landholdings and a low income (Watson, 1983). The system involves planting double thick rows of the tree legume ipil-ipil on cultivated contour lines with the aim of providing vegetative cover while the tree roots bind the soil. The leaves are pruned and used to provide mulch and fertiliser (*ibid*). Crops are grown in strips between the rows of ipil-ipil and non-perennial crops are rotated. Because ipil-ipil is a legume, *rhizobium* bacteria in its roots are able to fix nitrogen in the soil, further enhancing its fertilising abilities. Attempts have been made to introduce the SALT system at many sites in the Philippines, but although it is well known in the Visayas it is not widely used (Parrilla *et al*, 1988). However, in the Villaba area of Leyte it has been accepted by some farmers due to the influence of the Farming Systems Development Project-Eastern Visayas (FSDP-EV) (Lightfoot *et al*, 1985).

Despite the enthusiasm with which SALT has been marketed to the uplands farmers, there are some questions about its efficacy. For example, Parrilla *et al* (1988) state:

> While the SALT technology has been known for many years in the Philippines, and elsewhere in the wet tropics, a series of questions remains as to its long term technical viability. In particular, there are fears that the plants will then rot in the contour bunds. The consequence of this could, of course, be disastrous. It has been suggested that fears relating to its technical effectiveness have reduced the acceptability of SALT to hillside farmers (p. 2).

It has become apparent too, that many of the assumptions made by the assistance organisations have been faulty and oversimplistic, and that the barriers to adoption of soil conservation are not as simple as has been assumed. The assumption that cropping patterns and farm size were relatively homogenous, and that income was not greatly variable within a barangay have been shown to be wrong (*ibid*; Belsky, 1984; Lumley, 1997).

Non-Adoption of Soil Conservation Techniques

As mentioned above, the non-adoption of known soil conservation technologies has been observed on Leyte (Parrilla *et al*, 1988). One does not have to look far to see that this phenomenon is not confined to Leyte (e.g. Blaikie, 1985; Cramb, 2000) and it demonstrates the relevance of a case study in the Philippines to other countries. For example, in 1984 farmers were surveyed to assess the rate of adoption of conservation tillage methods in the northern grain growing districts of Victoria, Australia (Hurley *et al*, 1987). It was found that about fifty % of farmers claimed to have adopted such techniques (*ibid*). One of the goals of that research was to try and understand farmer decision making processes, and to find a broader explanation than financial incentive for the adoption or otherwise of soil conservation methods. While the researchers did not attempt a thorough socio-economic analysis, they assessed factors such as tradition, stewardship, and peer pressure, technology, physical environment, available physical resources, information and finance (O'Brien, 1987). O'Brien states:

> The study showed that economics [finance] is so overiding a factor that it becomes the main objective of the year's work. This economic objective is aimed at trying to balance the books, of trying to earn sufficient income to pay living expenses, operating costs and the oppressive interest bill ...

and that '... presentation of the 'right' and 'relevant' information is necessary but may not be sufficient to change a farmer's behaviour' (p. 2).

It is probable that these statements apply to farmers the world over and that farmer decision making is subject to similar influences wherever farmers are. For this reason it is assumed for the moment that circumstances cited in the the above quote apply to the farmers of Leyte, and that the results of the socio-economic analysis of the farmers in Leyte presented in this book will, in a general sense, apply to the wheat farmers in northern Victoria. It might be further assumed that both scenarios, which consider motives for making decisions, generally apply to farmers worldwide. It is clear that developed and less-developed countries alike have significant land degradation problems. Australia's massive losses of arable land to salinity are a good example (Pannell, 2000). Here, significant research effort has been made in developing technologies to combat salinity, including various agroforestry techniques (Race and Robins, 1998; Black *et al*, 2000). Yet as in the Philippines there is insufficient adoption of such technologies by farmers.

Philippines farmers are, of course, subject to greater extremes of nature than farmers in developed countries. Most live closer to the edge of insufficiency in meeting living expenses, and most face far more oppressive interest bills than Australian farmers. However their human motives for decision making are probably very similar. The relevance of economics, in its theory and application, especially in terms of the 'optimal' allocation of resources and the sustainability of current agricultural practices is likely to be consistent globally. More attention should be paid to translating research findings on motivating behaviour between regions and countries. For this reason the socio-economic survey in Leyte, Philippines can be viewed as being applicable to natural resource use generally. In addition, the results of this reasearch are seen as relevant to social and environmental policy with respect to farmer decision making, economic theory and sustainability, world wide.

Finding Reasons for Non-Adoption of Soil Erosion Solutions

In 1984 the Australian Centre for International Agricultural Research (ACIAR) held a workshop to discuss soil erosion management, at Los Banos in the Philippines. One of the issues discussed was the constraints on the adoption on soil conservation technology. One speaker suggested that the farmer would gain economically if he or she were to adopt soil conservation technology and thus the constraint on adoption was probably not economic (Velasco, 1984). This assessment suggests an oversimplified approach to analysing such problems and raised possibilities for new research. Another participant (Librero, 1985) tended to see the whole issue as one of socio-economics and recognised the links between soil erosion and the many social and practical difficulties of a farmer's life.

In 1986 ACIAR began funding a project to be conducted jointly by the Visayas State College of Agriculture (ViSCA) in Leyte, Philippines, La Trobe University in Victoria, Australia and the Department of Conservation, Forests and Lands in Victoria. The goal of the project was to identify the socio-economic constraints on the adoption of improved cropping methods by upland farmers in Leyte.

Parrilla (1992) used the socio-economic data base generated by this project as the foundation for her PhD analysis and concluded that it is beneficial for farmers to adopt soil conservation practices and that 'The socio-economic characteristics of the farm household constrains adoption and the use of soil conservation practices in the upland areas (of the Philippines)' (p. xiv).

One of the goals of the analysis in this book is to use the uplands in Leyte as a case study to examine the factors that influence decision making in adopting soil conservation measures. The findings will be used to propose alternatives and to explain where, and why, the gaps between the theory and practice of economically efficient allocation of natural resources arise.

Some of the socio-economic data used in this analysis were generated by ACIAR research project number 8541: The Socio-economic Constraints on the Adoption of Improved Cropping Methods by Upland Farmers in Leyte, Philippines. The original objectives of the study were as follows:

- To present a profile of hillside farming in Leyte, Philippines;
- To describe farm management with reference to production decisions and soil erosion prevention methods adopted;
- To describe the soil erosion prevention measures adopted in hillside farming;
- To determine constraints on the adoption of soil erosion prevention measures;
- To identify the parameters associated with the non-adoption of soil erosion prevention techniques; and,
- To consequently identify farmers who are likely to adopt soil erosion prevention measures currently available and to identify areas where further research is needed by soil scientists and others to overcome constraints which have been identified (Parrilla *et al*, 1988, p. 6).

The hypothesis on which the research was founded emphasised that if farmers had failed to adopt known and available technologies, then it was for reasons that the farmers considered sound. In other word the farmers were rational in making their decisions. Stent (1988) states that:

> Given this basic hypothesis, the adoption of a farming technology requires that:
> 1. The farmer knows about the technology.
> 2. The technology is consistent with the farmer's goals, and,
> 3. It is possible for the farmer to adopt the technology.
> Each of these conditions is necessary for the adoption of a new technology, none is sufficient (p. 2).

Having determined the hypothesis on which the research for the original ACIAR project was to be based, it was decided that the best way in which to test it would be by detailed survey of farm households over a period of eighteen months. This time period would ensure that data for an entire cropping year would be collected.

There are two provinces on the island of Leyte; Leyte and Southern Leyte. Three villages were selected from the province of Leyte and one from Southern Leyte. Each village was chosen on the basis of particular problems and characteristics required for the study. These included the potential for co-operation between researchers, farmers, local institutions and leaders; the presence of slope erosion; accessibility to research facilities; and representativeness of the area (Parrilla *et al*, 1988). Other factors also had to be taken into account. For example Stent (1988), when discussing methodology development comments:

> Rapid rural appraisal, though relatively cheap, often fails to identify the great inequalities within the agricultural sector and for that reason can be a most unsatisfactory method. The FSDP-EV baseline studies were of that type, and their inadequacy was recognised by the Mid-Term Evaluation team. Despite that and because of time constraints, they recommended that formal surveys could be conducted only if it could be established that the research objectives of the project could not be established by informal surveys. The manifest failure of farmers to adopt the FSDP-EV recommended packages [e.g. SALT] should however have suggested to them that already there was immediate need for a thorough and carefully designed household survey (p. 8).

Such household and village studies are not unprecedented. In 1988 (when the Leyte study was underway) a report on a major study of village life, conducted in India, was released (Walker and Ryan, 1988). This study, undertaken in India's semi-arid tropics had its genesis in the perceived need for '... quick answers to diagnostic questions to ensure accurate problem identification and subsequent technology design' (p. 2).

One of the goals of the research in India was to provide a broad base for identifying constraints to agricultural development. The issues assessed in the village studies were: Income; consumption and wealth; labour; land resources and tenancy; village markets; nutritional status and risk. *A significant finding of the study was that debt and a lack of access to formal credit had a strong impact on the cultural practices of the farmer.*

Research Objectives

The broad aim of this work is to examine why farmers may fail to adopt soil-conserving agricultural techniques even when they know that their present practices degrade their farmlands. The study focuses on farmers in upland regions of Leyte, an island located near the geographical centre of the Philippines. These farmers subsist from tilling deforested hillsides on the fringe of Leyte's mountain ranges. Leyte receives high rainfall in its wetter season and is exposed to tropical cyclones. The island's denuded upland slopes are unstable and in these conditions experience rapid soil migration and, from time to time, devastating landslides. The farmers in these regions generally have a good knowledge of soil conserving techniques; however most of them have decided not to adopt these techniques.

In seeking to explain these farmers' decisions, this book attempts to look beyond narrowly financial criteria and to examine also historical, social and economic factors that might have influenced farmers' assessments of the wider costs and benefits of adopting appropriate soil conservation technology. This book thus includes a brief socio-economic history of the Philippines. It also includes a chapter describing the geography of the archipelago, and in particular the geography of the Eastern Visayas island group in which Leyte is located. Assessment of the upland farmers' socio-economic status is informed principally by a detailed socio-economic survey conducted at four upland farming communities on Leyte over the years 1986-87, by staff from the Visayas State College of Agriculture (ViSCA), La Trobe University, and the then Victorian Department of Conservation and Environment, with funding and staff assistance from the Australian Centre for International Agricultural Research (ACIAR). Additionally the research draws on a further survey administered in 1993 to an identical sample of farmers. This survey was designed to provide data from which farmers' private time preference rates could be inferred.

The new survey was developed, in part, to test the hypothesis that farmers would not express different private time preference rates for money borrowed for different purposes, a hypothesis which is in keeping with assumptions in standard economic theory. If this hypothesis were to be rejected, then the knowledge of farmers' private time preference rates might help to explain their decisions concerning investment in soil conservation technology.

The earlier, far-reaching, examination of socio-economic factors influencing the adoption of soil conservation by Parrilla (1992) had accepted the standard economic assumption that interest rate paid on debt could be used as a shadow for the individual's discount rate. That assumption is also tested empirically in this book.

Research presented here assesses income and credit arrangements available to farmers at the study sites and analyses the relationship between the observed rate of interest charged on debt for a number of borrowing scenarios and the perceived discount rates of farmers for a similar range of scenarios, in the context of soil conservation adoption and sustainable resource use. Drawing on the results of the analysis, policies that would encourage more sustainable land use practices are proposed.

Theories and Research Goals

Soil erosion is a serious problem in developed and developing countries. The possible reasons for the non-adoption of soil conservation practices are examined, particularly in the light of economic theory and the individual discount rates of farmers. The Philippines has been used as the case study location because socio-economically it is representative of many developing countries, and because known soil conservation technologies, so often not adopted there, have international application. Originally, before Spanish colonisation (see Chapter 2), agricultural practices in the Philippines would have been mostly sustainable. However, historical events have led to unsustainable land use, mainly as a result of the commercialisation of agriculture, the development of world markets for agricultural products, the introduction of inequitable tenure systems and the lack of fair credit.

One of the main goals of this book is to use the results of socio-economic surveys in Leyte as a case study. The most important part of the analysis concerns the effect of various socio-economic factors on farmers' individual discount rates, and of discount rates on adoption practices. This analysis was conducted in the broad context of assessing the economic aspects of sustainable use of natural resources globally. Many governments are developing policies and strategies aimed at using their natural resources efficiently. There have also been precedents for international co-operation in trying to find a solution to environmental problems (for example the Earth Summit in Rio, in June 1992). However, much of the work so far has been aimed at fulfilling goals for climate change and greenhouse gas targets. While these goals are both worthwhile, and related to other environmental issies considered here, in formulating policy, there does not seem to have been sufficient attention paid to research into the socio-economic causes of environmental degradation and the over-use of resources.

There is growing concern about environmental quality and people are beginning to realise that far from being just a question of ideology or aesthetic values, environmental degradation is imposing large economic and social costs world wide. In some cases, especially in less developed countries, environmental damage is threatening the life and livelihood of millions of people directly dependent on the land, and on access to goods and services such as clean water and adequate food supplies (Palo and Mery, 1996; Sponsel *et al*, 1996; Pearce, 1999; Japan Environmental Council, 2000). Despite sophisticated technical knowledge and expensive aid programs, land users in developed and less developed countries are making decisions which superficially appear to be irrational. For example technologies which in the long run would improve their living standards are not adopted.

Economic theory has much to say about the optimal allocation of resources, about efficiency and about rational behaviour. However, what may seem simple and rational to theorists and policy makers may not seem rational to people making decisions about their day to day lives. Economics is an inexact discipline, both as a science and as a social science. There are many branches of the discipline and, unfortunately, particularly lately, economics has become polarised, politically and ideologically. Thus policy development in academic, bureacratic and government circles may often have more to do with ideology and politics than with problem solving (for example see Pearce *et al*, 1990; Lumley, 1999; Pearce, 1999).

Partly because of the way it is reported in the media and partly because of short term government goals, economics has come to be almost synonomous with finance and commerce. Essentially, neo-classical economics purports to be about the distribution of scarce resources amongst competing ends, and it really has little to do with money *per se*. Lumley (1991) states that:

> Many economists feel the need to clear up confusion arising from various parties, including vested interests on both sides of the equation [of natural resource use], deliberately or accidently misleading society about what is 'economic' and what is not. Generally when people refer to 'economic' they mean financially viable or profitable to a small sector of society. The overall discipline has much less to do with money, and much more to do with the efficient allocation of resources, and with increasing the well-being (welfare or utility) of society as a whole (p. 7).

In this research a detailed assessment is made of the various influences that impinge on farmer decision making. This cannot be done by choosing one factor out of the overall context. For example, it is clear that a knowledge of feasible technology is not enough to persuade a farmer to adopt it, nor is the existence of private property rights or a buoyant market sufficient for 'rational' decision making. For this reason the results of the household surveys in Leyte are assessed in the context of the social, cultural, political and economic history of the country and in the global context.

The role of economic theory is also important in analysing the apparent failure of decision makers to allocate resources efficiently because such failure suggests that there is something wrong either with economic theory, or with the way it is interpreted and applied.

Some theoretical issues affecting practical outcomes which influence the understanding, development and interpretation of economic theory, and the roles of financial and socio-economic analysis are as follows:

- The practice and significance of discounting and the choice of discount rate;
- The significance of property rights and land tenure on natural resource use;
- The role of open capital markets and the theory and assumptions relating to them;
- The role of politics and ideology in the choice of economic policy tools by governments and industries

The results of this analysis help to identify whether land and environmental degradation is caused by the 'non-optimal allocation of resources', and if so, what factors lead to such an allocation. There will also be some consideration of the way in which economics can be applied and interpreted to assist with a more desirable allocation of resources, in a manner leading to long term sustainable use of resources, and a more equitable distribution of wealth globally.

Outline of this Book

The objectives of this study, a review of ideas in sustainable development, and an overview of land degradation issues, have been presented in Chapter 1. Chapter 2 comprises a brief social and economic history of the Philippines. This chapter provides a perspective on the current systems of land use and land tenure, and on the development of export markets and subsistence agriculture now prevalent in the Philippines and many other less developed countries. The material is presented chronologically to show how Spanish colonialism initiated a process which catalysed change from sustainable agricultural systems, where private land ownership and cash cropping were unknown and world markets did not exist, to a system of private land ownership, private tenancy, and plantation agriculture largely dependent on protected markets under American colonialism. These changes led to the present system of high risk, marginal subsistence agriculture on unstable slopes practised by the majority of farmers in this case study.

Chapter 3 provides background for the socio-economic analysis of factors influencing decision making covered later in the book. It describes the reality of life for the farmers and the risks associated with the weather conditions, harvest failure, price volatility, food availability and access to social services. Chapter 3 contains a description of geophysical conditions in the Philippines and Leyte. Included in this is a brief outline of natural disasters with a description of the consequences of severe upland degradation under the cyclonic conditions which often occur in countries like the Philippines. This gives a good understanding of the sort of risks and tradeoffs which farmers must face when making decisions about obtaining and paying for credit, and implementing soil conservation, and provides useful background material to Chapter 4.

In Chapter 4 a full description of project establishment and survey development is provided. This chapter outlines the process of site selection and gives a brief description of each of the study sites, which consolidates the information on Leyte and the Philippines presented in Chapter 3, and gives further insight into living conditions for the farmers. This is followed by a discussion of the questionnaires used in both the surveys, and an assessment of related studies conducted elsewhere.

Chapter 5 is a theoretical chapter which provides a context for the analytical work on interest rates paid on debt at the study sites, and their relationship to the private time preference rates of farmers and hence the farmers' perceived discount rates. Chapter 5 provides the theoretical link between the background Chapters 1 to 4 and the analytical Chapters 6 to 8. In Chapter 5 a theoretical economic discussion about the role of interest and discounting is presented. Some standard theoretical assumptions are examined and the role of time preference rates is discussed.

Chapter 6 is devoted to preliminary analysis of the data generated from the surveys. Each site is analysed separately and differences in behaviour and conditions between the sites are assessed, illustrating the non-homogeneity of the sites. Income and income inequality are examined in the light of tenure status and interest paid on outstanding debt. This is very important in view of the later analysis of discount rates and willingness-to-pay questions assessed in Chapters 7 and 8.

In Chapter 7 a series of hypotheses about interest and discount rates is tested. It is shown that there is no significant relationship between mean interest paid on debt and mean private time preference rates of farmers at any of the study sites. This chapter is important because of the testing and interpretation of results relating to interest rates and perceived discount rates and soil conservation adoption, and because of the relationship of these findings to assumptions in standard economic theory.

In Chapter 8 there is some further hypothesis-testing and the overall results are interpreted. A discussion about the implications for policy development is presented. Chapter 9 is the concluding chapter and following a summary of the results, proposals are made for further research and for policy implementation. The contribution of this book to the global sustainable development debate is discussed.

Chapter 2

A Brief Social and Economic History of the Philippines

Introduction

The Philippines has experienced one culture being laid upon another for more than 1000 years. It is extremely difficult to identify any single factor which has led the country to where it is now, yet the layers of culture hold clues to why most Leyte hill farmers do not adopt known soil conservation technologies. When the Philippines was settled, land-use patterns were sustainable in the long term. Systems of government and agriculture introduced to the country since Spanish colonisation have led to unsustainable land use and a socio-economic culture that makes progress towards sustainable resource use difficult.

This chapter outlines the progression from usufruct land use systems with no formal private ownership and small scattered populations, to the predominance of a national government overseeing protected international cash-crop markets. The crops are mainly produced on tenanted private lands, where crop cultivation to feed the population is secondary to the role of cash generation for the land owners.

Before Spanish Settlement

The original inhabitants of the Philippines archipelago were known by the Spaniards as the Negritos, who are believed to have migrated over land bridges during the last glacial period, which ended about 10,000 years ago. There were waves of migration from other Southeast Asian countries after those land bridges had disappeared. These later immigrants included the Malays, the Proto-Malays, the Indonesian Malays, and Sinitic and Chinese peoples. These waves of immigration took place over thousands of years and continued beyond the era of Spanish colonisation (Seekins, 1984).

There is archaelogical evidence in the Philippines which places human settlement at more than 45,000 years ago. In the Tabon Caves of Palawan Island burial grounds containing pottery, metal burial urns, tools and jewellery of glass, jade, iron and copper have been excavated. These demonstrate evidence of continued occupation from seven and a half BC to the fourteenth century AD. From the tenth century AD many of the artefacts were of Chinese style or origin (Fox,1970).

In pre-Spanish times there was little settled agriculture with the exception of the permanent field rice farmers of Northern Luzon, the Ifugao tribe. These people established the first agricultural settlements and grew millet, then rice, on a complex system of terraces in the hilly districts of the island. The crops were irrigated and the people thus escaped the risks of flood and drought which occurred from time to time in the lowlands (Wernstedt and Spencer, 1967).

In the southern islands, such as Sulu, the people were coastal dwellers who had a good deal of contact with commodity traders, fisherman, slave traders and pirates from other Asian nations. They relied on fishing and trading and had little need for a sedentary lifestyle. Indeed many of the coastal dwellers were known for their skills in navigation and their ability to build ocean-going boats. They were also skilled engineers and architects (McCoy and de Jesus, 1982).

There were other lifestyles in the archipelago. A typical variant was the coastal village settlement, which had access to the sea for fish and other ocean goods, and access to the inland forests for cropping and food gathering. These people employed shifting cultivation for the plants they wished to grow. There were also people with established villages who lived in a similar manner to those in the coastal settlements, but relied on the swamps and the rivers, rather than the sea, for aquatic harvests (Seekins, 1984). As new immigrants arrived over the centuries, these parallel living patterns were built upon and varied, but the essential elements of the socio-economic systems remained.

The island environments varied, and living patterns were adapted according to the natural resources available. The Philippines had a wealth of native plants, and those such as abaca, calamondin and lanzon were adopted for cultivation. In addition, immigrants brought with them the plants that were cultivated in their countries of origin. These included millet, rice, citrus fruits, bamboo and bananas. Many tools and technologies were brought from the Asian mainland (Wernstedt and Spencer, 1967).

There was also a political pattern which evolved according to the origins of the immigrants. The Moslems, who arrived about 130 years before the Christian Spaniards, introduced the notion of regional rulers, such as *sultans*, and local chiefs, such as *datus*. Prior to Moslem settlement the concept of territory was unknown, except to the farmers of Luzon (whose cultural patterns do not appear to have been adopted elsewhere).

The style of settlement adopted by most of the islanders took the form of the *barangay*. The term was introduced by settlers from Indonesia and Malaysia, and replaced *barrio*. Both terms are used in the Philippines today, and although they represent what Westerners would think of as a village, these communities have a strong parallel with the early settlements.

There was a social hierarchy in the barangays which was based on kinship ties. At the apex of this was the Chief, who belonged to the class of nobles. Below the Chief came the freemen, and then the dependents.[1] The lowest stratum of people in the barangay, the dependents, fell into several sub-groups; these were the workers, the ex-freemen (who had been downgraded due to socio-economic transgressions involving crime or debt), and slaves (who were usually war captives).

These localised social units, which involved a low level of socio-political institutionalism, together with the nature of the geography, topography and ecology of the archipelago, meant that there was a good deal of regional autonomy and inter-regional hostility, which later assisted the Spaniards with their conquest of the islands.

At the time of Spanish settlement the islands were sparsely populated, with a total of only about 500,000 inhabitants (Saastamoinen, 1996; Seekins, 1984). Trade between the Philippines and other nationals, including Hindus, Indonesians, Chinese, Japanese and possibly Arabs (who brought Islam to the Asia), had been conducted for some time.[2] The Chinese were trading with the islands in the 10th century and the Japanese in the 15th century. Both of these groups established trading settlements.

Islam was brought to the archipelago from Indonesia. It was well established in the Sulu area by the 16th century and had spread to Mindanao and from there to the Manila region of Luzon by 1565. The idea of territories ruled by a sultan who had sovereign rights over the Barangay chief, although introduced by the Moslems, did not spread outside discrete established boundaries. In 1521, when Spanish colonisation began, most of the population lived in the established barangay settlements. (Seekins, 1984; Community Aid Abroad, 1984; Wernstedt and Spencer, 1967; Foreman, 1899; McCoy and de Jesus, 1982).

At the time of Spanish settlement the inhabitants of the Philippines had a very rich culture derived and adapted from a number of sources over a long period of time. The dominant influence had been the Malay and Indonesian cultures, the members of which which were often sophisticated and well educated. According to CAA the people could read and write using leaves, bark and bamboo. Their cultural heritage included stories, songs, fables, and criminal and religious codes.[3] A Spanish chronicler, Pedro Chirino, wrote: All these islanders are much given to reading and writing, and there is hardly a man, and much less a woman, who does not read and write (de Jesus-Viardo, in Community Aid Abroad, 1984, p. 8).

Spanish Colonisation

In the 15th century Spain and Portugal were feuding over rights to colonisation and trade. Pope Alexander VI intervened to settle the question of demarcation. In an attempt to resolve the feud he issued a Papal Bull dividing the world into an Eastern hemisphere and a Western hemisphere. The Eastern hemisphere was to fall under the rights of the Portuguese to colonise and trade, and the Western hemisphere was for the Spaniards. This arrangement was adopted in the Treaty of Tordesillas. However, argument between the Spaniards and Portuguese continued, partly because of the wilfully lax interpretation of the Treaty by those party to it. Although Spain later claimed the Philippines, the islands were in the area allocated to Portugal (Lumley, 1998 (c); Foreman, 1899).

Magellan was responsible for the first Spanish claim to the Philippines.[4] Magellan's first expedition sailed in August 1519 and landed in Rio de Janiero the following December. It sailed through the since-named Magellan Straits along the coast of Patagonia, and reached the Pacific Ocean in November 1520. After crossing the ocean the fleet coasted along the north of the Philippines island of Mindanao in the south of the archipelago. The first Roman Catholic mass was celebrated on the island of Leyte (Community Aid Abroad, 1984). The territory was then claimed in the name of Charles I of Spain. For these islands, as yet unnamed by the Spaniards, this was the beginning of more than 350 years as a colony of Spain.

Magellan and his fleet sailed on to the island of Cebu, where they were greeted by an angry crowd of islanders who were amenable neither to the idea of the Spaniards landing nor to conversion to Catholicism. After discussion between the Cebuanos and a Batuan chieftain who was with Magellan, who gave them assurances of the Spaniards' peaceful intent, a *Pacto de Sangro* (blood pact) was struck, and the island's leaders were, according to Foreman (1899), subsequently baptised.

In the 16th century there was little solidarity between the islands (Seekins, 1984) as they were ruled not centrally but by local tribal leaders (barangay chiefs), and the chiefs of Cebu and Mactan were at odds with each other (Foreman, 1899). The chief of Mactan, Lapu Lapu, was displeased by Magellan's attempts to convert his barangay to Catholicism and he had Magellan killed with a poisoned arrow in April 1521 (Lumley, 1998 (c); Goodno, 1991). In August 1543 another expedition was sent to the islands and the archipelago was named the Philippine islands in honour of Charles I's son, Philip. This Philip ascended the throne of Castile when his father abdicated in 1555.

Animosity between Spain and Portugal continued. In 1558 the quest for the spice islands was being maintained and the value of the Philippines to Spain was still unclear. In view of the desire of Spain to possess the spice islands, many Spaniards pressed the Viceroy of Mexico, of whom the Philippines was a protectorate, to annex New Guinea instead of the Philippines - this did not happen, and an expedition to the Philippines departed Mexico in November 1564. This expedition was led by the Basque navigator Miguel Lopez de Legaspi. The general aim of this journey was to conquer and Christianise the Philippines Archipelago. Having put into port at Cebu in the Visayas, Legaspi sailed to Mindanao where he was told of the wealth and power of Cebu. Legaspi decided to claim Cebu for the King of Castile. Again the inhabitants of the island strongly resisted the Spaniards' attempts to colonise them but after a fairly lengthy occupation the islands were declared to be under Spanish control and their occupants were named as Spanish citizens (Foreman, 1899; Goodno, 1991).

The early history of Spanish colonisation is not clearly chronicled, and some contemporary accounts are in conflict. However in 1570 Spain set out to claim the large island of Luzon (on which Manila is situated) and after opposition from the islanders this aim was acheived. Mindoro island was claimed soon after. On 24 June 1571 the City Council of Manila (previously known as Maynila) was constituted.

Soon after Spain had established its claim to the conquered territories it established a system of government that while zealous in its missionary work did not have an overwhelming impact on the use of the country's natural resources or on its systems of husbandry and food collection. Land use continued to be sustainable. Foreman (1899) states:

> The system ... was to let the conquered lands be governed by the native caciques and their male successors so long as they did so in the name of the King of Castile. Territorial possession seems to have been the chief aim of the invaders, and records of having improved the condition of the people or of having opened up means of communication and traffic as they went on conquering, or even having explored the natural resources of the colony for their own benefit, are extremely rare (p. 34).

This situation continued, more or less unaltered until the colonies were opened up for international trade in the late 18th century. Thus, as discussed later, it was the development of international trade, which was directly associated with the beginnings of plantation agriculture for cash crops, that led to the beginnings of unsustainable land use (Uitamo, 1996; Lumley, 1998 (c)).

The Rise of World Trade

There was little large-scale trade between the Philippines' peoples and their European colonisers in the first centuries of Spanish occupation. The Spaniards did not discover precious metals or minerals which they could exploit, nor did they discover the spices which they had originally sought. In fact the colony was unprofitable for Spain and its treasury was nearly bankrupted as a result of the long 17th-century war with the Dutch and the ongoing conflict with the Moslems from the south (Seekins, 1984).

In the first part of the 17th century Manila enjoyed a brief ascendency in the world silk trade, re-exporting silks from China to Europe, via Mexico. Trade peaked in 1605, declined after 1650, and did not rise again until the end of the 18th century (McCoy *et al*, 1982). In the interval the islands gave satisfaction only to the missionaries, and Manila survived as a colonial outpost by virtue of its encircling haciendas (the Friar estates).

In the late 18th century, Europe's world trade increased and Spain turned to its colonial outpost for participation. Shipbuilding techniques had improved since the earlier trading days and industrial growth had provided the incentive to make the Philippines more accessible to European trade, which accelerated in the 19th century. It was this increasingly profitable trade which gave the stimulus to start exploiting the natural resource base of the islands. This base had been only marginally touched during the previous years of Spanish rule.

The resultant socio-economic change had parallels all over the world in countries where colonial rulers have attempted to extract a return from their territories. McCoy writes that as the Philippines developed its first new links to intra-Asian, European and North American markets:

> Between 1780 and 1920, roughly speaking, most of the archipelago experienced a socio-economic transformation ... The clearing of the Philippine frontiers was high social drama complete with sudden migrations, land grabbing and aboriginal slaughters that accompanied the process in America, Australia or the Argentine. Migrating outward from the crowded Ilocos coast of northwestern Luzon, peasant pioneers cleared

the Central Luzon Plain and the Cagayan Valley. Quitting a similarly parched and overcrowded ecosystem in western Panay, Ilongo pioneers cleared the rainforests of Negros Island for vast sugar plantations. By the 1870s the archipelago's several regions produced a range of commercial crops for intra-Asian and international markets – tobacco in the Cagayan Valley of north-eastern Luzon, rice and sugar in Central Luzon, abaca hemp for sailing cordage in the Bikol Peninsula, sugar in the Western Visayas, and Chinese foodstuffs in Sulu (McCoy and de Jesus, 1982, pp. 6, 8).

It appears that this new international trade was the catalyst for establishing the socio-economic system and hierarchy which are dominant in the Philippines today, but that these remained firmly based in the system of hereditary rule put in place by the Spaniards and entrenched in the dormant years (from Spain's perspective) of colonial mediocrity. Thus a Western economic system was introduced relatively early by a hierarchy imposed from without, a hierarchy which probably allocated resources away from an internal economic or social optimum. McCoy and de Jesus (1982) state:

> Whereas in Java traditional production systems were exploited or preserved, in the Philippines production was more commercialised, or capitalist, from the outset. Emerging regional elites consequently played a critical entrepreneurial role in the growth of Philippines exports and some local peasantries began to operate as rural wage labourers. The archipelago thus became the first area in Southeast Asia to develop indigenous commercial elites employing modern production methods and a rural wage labour market (*ibid*).

However, during the period of trade development the different regions of the archipelago operated different labour systems, and the system of wage labour was by no means uniform across the islands. In central Luzon the plantation owners used a tenancy system, in Sulu, there was a slave labour system, and the Chinese planters of Negros, while initially using tenancy, passed onto wage labour and debt bondage (which we still see in the Philippines today). In the highlands areas the people were marginal to the trading operations and were involved in cash cropping activities (*ibid*).

In 1762 Spain had allied itself with France in the Seven Years War, which was fought from 1756 to 1763. The British had been trading around the islands for some time and in October 1762 the British East India Company captured Manila as a direct action in the war with Spain (Seekins, 1984). Prior to this the British had taken a galleon, the *Philipino*, which had been sailing from Mexico to Manila. Part of its cargo had been a subsidy from Mexico to support the colonial administration, for at this stage the country was a Mexican dependency (Foreman, 1899). Meanwhile, since Spanish settlement there had been a number of rebellions and uprisings throughout the archipelago by the local people, who resented paying tributes to Spain.

The socio-economic transformation of the archipelago continued into the 19th century. In the 18th century Britain had become very active in the trade between Asian countries. The British dealt mainly with the Chinese mestizo merchants, and they imported and exported cloth and local textiles. Some writers (e.g. Seekins) have inferred a link between the significant early textile industry and the transition to agricultural exports in the middle of the 19th century.

Early in the 19th century many of Spain's South American colonies were gaining independence, including Mexico in 1821. This deprived Manila of its subsidies and closed the protected South American markets, forcing the Philippines to develop a more financially self-sufficient economy. By this time merchants from a number of nations had come to Manila and they assisted with integration in international markets which supplied Asian raw materials to the manufacturing industries of Europe and North America. In 1834 Manila's port was opened to free international trade (Seekins, 1984; Goodno, 1991).

The Development of Agriculture

As described earlier, established patterns of settlement meant that the lands around a barangay were used according to need and were commonly held by the settlement itself. Often there was little cropping, and the fruits, plants and other foods required for consumption were gathered from the forest and from the river or the sea.

Prior to Spanish colonisation, land in the islands was defined in terms of how it was used, rather than in terms of who owned it. Wernstedt and Spencer (1967) state that:

> Ownership was vested in a communal group, meaning a society, a people, a clan, or some other organisational entity among a given population. Ownership went with occupancy of a territory by a population unit. The tenure of land was by usufruct, meaning the right to use land for any of several purposes but not the right to alienate that land in any way. Tenure operated for several purposes in specific ways according to the social structure of the particular group. Among socially stratified peoples or tribes there were classes or persons at the lower extreme who were

excluded from particular rights of tenure, whereas at the upper extreme there were classes or persons who could demand tribute in labour, service or a share of the yield produced by the exercise of particular tenure rights from either the whole population or given classes below them. Between these two extremes were varied patterns by which most of the small population held tenure rights for any purpose and to retain the products thereof. Shifting cultivation, hunting, food gathering, appropriation of forest and mineral resources and fishing all operated within this general framework of ownership by a group and tenure by usufruct. Among the north Luzon highlanders modifications of this traditional pattern approached concepts of private ownership in the limited sense that tenure applicable to permanently cropped terraced lands was held during the continued occupance of any given field (p. 187).

When the Spaniards arrived, the most common approach to land use was a system of shifting cultivation known as *kaingin*, practised by people who became known as *Kaingineros*. This system is sometimes referred to as slash-and-burn agriculture, but was very different from today's 'slash and burn'. A small area of forest was cleared and the crops to be grown were planted in the clearing. These crops consisted of rice, bananas and yams, with less-used crops such as coconut, sugarcane and abaca (Manila hemp) to complement the major crops.[5] Other foods were gathered from the forests. When the crops had been harvested the land was left to be reclaimed by the forest, and was not cropped again for a number of years.

After establishing a centralised government in Manila, and regional jurisdictions in the hands of a local elite which built upon the existing systems of the main Malay ethnic groups, the Spaniards developed Crown-land grants known as *encomiendas*. These entitled their holders to levy labour service and tributes of goods from the resident population. These grants were taken up by Spanish soldiers and bureaucrats, the local indigenous elite and the church. The Spaniards, really only understanding a system of private land rights, recognised native ownership only of land that was actually under cultivation, or occupied. All the unoccupied land of the Philippines, most of which was used by the resident population in common ownership, became Crown land which the Spaniards entitled themselves to dispose of. This was the beginning of private land ownership in the archipelago. While this system of grants went into decline in the early 17th century, land control was now included firmly within the concept of private ownership. Eventually the local elite and the educated Spanish and the church were able to claim title to the land, and its control was removed from the ethnic communities and placed in the hands of the title holders (*ibid*).[6]

Meanwhile, the Catholic Church continued to exert a strong influence in the Philippines. The friars of the church had spread into the regions and they were answerable directly to their regional heads, rather than to the authority of the Bishops in Manila. The orders of friars had been adept at taking landholdings and came to own large tracts of land under grants and then outright. Many of these land holdings had been given to them as donations to the church to develop churches, schools and other establishments, although the orders also squatted on native lands, and grew rich among resentment from the traditional owners of the land (Saastamoinen, 1996; Seekins, 1984). These friars came to hold massive estates in the islands, particularly in Luzon, and were among the earliest proponents of plantation agriculture. The friars let their lands to Chinese mestizos who grew export crops for cash.

The friars also controlled the cultural and intellectual life in their regions, because they were in sole charge of education. This monopoly could have been broken in 1863 when free state primary schools were established, but the friars were responsible for supervising the school system. The Catholic church appears in many ways to have been the *de facto* ruler of the Philippines' regions. According to some reports (*ibid*) the friars were corrupt, racist, and in violation of their vows, as well as being contemptuous of the native people. These, then, were the first extensive export agriculturalists of the islands.

The first plantations were cultivated for sugar, and the practice of producing agricultural commodities for export, a practice dominated by the haciendas in a system which, arguably, led to an unequal distribution of land, continues to this day (Community Aid Abroad, 1984). While sugar was the original cash crop and still has a significant role in the Philippines economy, more export cash crops were planted from the time that international trade was established. By the late 19th century these crops included tobacco (grown under government monopoly for local consumption in the late 18th to mid-19th centuries), rice, abaca, coffee, cocoa, coconut (also a significant export today), gum, and hardwoods (of increasing significance today). While the hardwoods were and are harvested from the native forests they have had and will have a major impact on agricultural production and natural resource use (Uitamo, 1996; Saastamoinen, 1996).

The export of agricultural products and, later, minerals, brought financial self-sufficiency to the islands, and set a pattern of socio-economics which has had an enduring influence on natural resource use and individual and collective lifestyle patterns. These will be analysed later in this thesis, in the hope of developing explanations for what may appear to be irrational decisions made by contemporary farmers in the Philippines.

The Americans

Over the years of Spanish colonisation there had been a number of rebellions and uprisings against Spanish rule. In the late 19th century opposition to Spain had become very strong and it was more organised than it had been earlier. Many of the wealthy Filipinos (*ilustrados*) had been educated in Europe and with the peasants they had developed a taste for nationalism. Some of the ilustrados published nationalist writings in Europe.

In the 1896-98 risings for independence from Spain, the landless and the educated elite joined forces, although the ilustrados gained control of the cause. In 1896 the Spaniards colonial government executed Jose Rizal, the revolutionary leader, writer and hero (Goodno, 1991).

In 1898, as a result of a disagreement over Cuba, the United States declared war against Spain and set out to destroy the Spanish fleet anchored in Manila Bay. A revolutionary Filipino leader, Aguinaldo, had held talks with the United States of America (U.S.) in late 1897, and the Philippines rebel movement viewed its relationship with the U.S. as a partnership against Spain. The Filipinos assisted the U.S. with military intelligence against Spain, although the U.S. would make no written commitment to Philippines independence.[7]

The Philippines was granted independence by the USA in 1946, after a period of Japanese occupation in World War II. The Americans attempted to prepare the nation for independence by introducing concepts of Western-style democracy in the education system. However since independence, national and regional leadership has generally fallen upon members of the traditional socio-economic elite, such as the Aquinos and the Marcoses (Bunge, 1984).

As would be expected, the USA had a strong influence on the Philippines economy during its period of rule. North America provided the main markets for agricultural products and determined, by way of demand, the choice of plantation land-use for cash crop production. Free trade between the countries was guaranteed until 1954 when American import tariffs began. Uy Eviota (1992) states of American colonialism that: ... 'the economic restructuring that had begun in the latter years of Spanish occupation quickened and the effects became more widely felt. The increased penetration of industrial capital had altered the production process and market relations' (p. 76).

Agreements between the two countries also entitled American investors equal rights to resource development and permitted operation of military bases, which remained until 1991 and were closed as an indirect result of the eruption of the volcano Mount Pinatubo.[8] One area little influenced by the Americans was land reform and land tenure, which owes much of its present form to the settlement activities of the Spaniards and the Catholic Church.

Post-Colonial History

During World War II the Japanese had invaded and occupied the Philippines. In September 1944 US General MacArthur persuaded President Roosevelt that it would be in the interest of the United States of America to invade and liberate the Philippines. To this end 200,000 American troops landed on Leyte, near the city of Tacloban, in October 1944. Quezon, the Commonwealth President of the Philippines, had died two months earlier and his successor, Sergio Osmena, accompanied MacArthur on his invasion. The islands of the Visayas were recaptured with the aim of advancing to recapture the island of Luzon and the city of Manila. This goal was realised in February 1945 (Goodno, 1991). The liberation of the Philippines was achieved in July 1945, but:

> The Americans, history says, came as 'liberators' and, to assure the Filipinos of this, the US kept to its promise of independence in 1946, but first it had to restore elite rule and protect what policy makers in Washington decided were US interests (*ibid*, p. 40).

The victorious Presidential candidate in the ensuing elections, Roxas, campaigned on the basis of 'Who was America's friend?' and 'Who would bring in more U.S. aid?'. Roxas defeated Osmena to become President of the Philippines, and the U.S. Congress passed a 'parity' amendment of the Bell Trade Act which granted special rights to US investors. The U.S. Bell Trade Act required that duty-free trade between the U.S. and the Philippines be conducted until 1954, and it set quotas on Philippines exports to the U.S. (full duties would not be realised until 1973). The parity clause of the agreement gave U.S. citizens the same rights to ownership of property, industry and public utilities in the Philippines as Filipinos had, but offered no reciprocal arrangement. The Bell Trade Act was adopted and U.S. rehabilitation funds to the Philippines were tied to the agreement (*ibid*). This scenario of regulated markets, U.S. intervention and a chosen elite set the scene for Philippines independence.

In 1947 the Military Bases Agreement was endorsed by the U.S. and Philippines governments. The major bases, the Navy's Subic Bay and the Air Force's Clark Field, and other sites, were proposed for 90-year leases and the Filipinos assented, at least in principle, to the developments (Friend, 1986).

It seems surprising that a country, having fought for so long for independence, and having won it after 400 years, should agree to being tied to such restrictive agreements balanced in favour of its former colonial master. Part of the explanation may lie in the power of *utang na loob* – the debt of honour which is owed to a patron and protector (*ibid*). This debt, which must be paid, has had a strong influence in the power base of Philippines politics and policy, through colonial history and independence to the present.

Meanwhile, there was internal unrest in the nation. A huge peasant organisation, the *Hukbo ng Bayan Laban sa Hapon* (or *Hukbalahap* or *Huk*), which was established from a range of smaller organisations in 1942 to fight the Japanese occupation, was causing problems to the new government. The organisations which made up the Huk were originally formed to unite tenant farmers against the hardships they suffered from their landlords. The Huk was rigid, sophisticated and thoroughly organised, and its aim was to wrest control of the land from the landlords (Davis, 1987). Another organsation with similar goals, the *PKM* (National Peasants Union), was backing the *DA* (Democratic Alliance) which had been successful in the elections in Luzon but had been excluded from Congress. The *PKP* (Communist Party of the Philippines) was also involved in pushing for land reform. The DA, PKM, PKP and the Huk held meetings with Roxas in an attempt to settle their grievances, but did not achieve their goals (Goodno, 1991). The problems caused by landlessness and poverty had been exacerbated by the fact that most of the $400 million of US postwar rehabilitation aid was used to profit the landed elite by paying for restoration of their private property (*ibid*). Goodno states that:

> Repressive taxes, unfair land tenancy, disenfranchisement, continued suppression of peasant organisations, economic crisis spurred by corruption and the refusal to grant the Huks the benefits given to other veterans of the anti Japanese war all helped to fuel unrest[9] (p. 42).

On 30 December 1965, Marcos ascended to the office of President of the democratic republic of the Philippines. For the first four years it remained a democratic republic. There was a two-party system, with executive power being vested in the President and legislative power being vested in the Congress. The Congress was composed of the Senate and the House of Representatives (Rodriguez, 1985). Ferdinand Marcos ruled the Philippines for 20 years. Much has been written of him and his family: Of their graft and corruption, their moral poverty and their lack of conscience in depriving an already poor people of the meagre wealth to which they were entitled.[10] There is room here only for a brief summary of those years. However, Marcos certainly has had an impact on the way in which natural (and other) resources were allocated in the Philippines, and he added to the complexities which determined the manner in which wealth was distributed and the way in which the landed elite, inherited from Spanish days and beyond, consolidated their power. Rodriguez (1985) states that: 'When President Marcos came to office, ... he naturally inherited the handicaps present in the system – the culture, the politics, the exploding population growth and the stagnant national economy' (p. 23).

Seagrave (1988) stresses the importance of the Chinese (who had an influence in the Philippines from the 10th century) over the years, and in two paragraphs he illustrates *his* views of the effect of Spanish and Chinese cultures on present day Filipinos:

> Most Spaniards, like the Chinese, came to the islands without women and made temporary arrangements with Malay girls, producing prodigous numbers of mestizo children. Fortunately, Chinese mestizo children were not considered Chinese. Raised as good Catholics by their Malay mothers, they could also come and go at will, own land, and engage in business, more or less as Malay Filipinos did. However, since they had access to Chinese credit, ... Chinese mestizos were in a much better position to buy property, and to act as middlemen or money lenders, which gave them exceptional leverage.
>
> Ordinary Malays ... naturally tried to emulate their Spanish rulers by throwing pig roasts on feast days, Christenings, confirmations, weddings, or any other occasion that came along. Without cash, in a rice and fish subsistence economy, they had to borrow money from the Chinese, using their traditional land as collateral. When the debt could not be paid the land was forfeited [this system still exists]. By this indirect form of extortion, more and more land came under the ownership of Chinese mestizos. The original Malay landowners became merely tenant farmers in their own country (p. 9).

Thus, early in the history of settled agriculture and private land ownership in the Philippines, access to cash and the nature of credit had an influence on land tenure and land use. This in turn has had an effect on the sustainability of agricultural practices.

In his early years Marcos tried to consolidate relations with the USA and focussed on development plans and efficient tax collection; he also greatly increased the military budget. However, by 1969 violence had escalated, relations with neighbours such as Malaysia had worsened and the economic system was in crisis (Noble, 1986).

It had not taken long for people to realise that graft and corruption had become worse. Rodriguez (1985) comments that the indiscriminate granting of credits and guarantees in the establishement of new industrial plants led to useless projects and a waste of foreign exchange and resources. This was evident in sugar centrals, cement plants and steel projects (*ibid*).[11]

By 1970 things were not going well with Marcos. Elections had been held in 1967 and 1969 and inflation was rising. Prices of basic commodities such as rice, fish, pork and poultry had risen sharply since 1965. There was a growing feeling of nationalism, and some of the ruling elite and the oligarchs were falling out with each other. Martial law was introduced in September 1972, the day after anti-Marcos demonstrations at which the car of defence Minister Juan Ponce Enrile was ambushed.[12]

The government continued to sabotage empty buildings and blame the destruction on its opponents (of whom it had many), including revolutionaries from groups such as the New Peoples Army (NPA), who appeared to have nothing to gain from such actions. By this time Benigno Aquino was speaking in open opposition to the government. There were continuing problems between the Christians and the Moslems of Mindanao and the Sulu Archipelago. The introduction of martial law severely restricted freedom of speech and freedom of the press. There was talk of developing a new democracy based on the old barangay system and, again, of land reforms. However proposed reforms did not cover landless labourers, small tenant farmers, or any tenant farmers not involved in rice or corn production (Noble, 1986). In 1974 inflation reached nearly 35% (see table 2.2) and real wages declined by about 20%. The change in real wages for skilled and unskilled workers was dramatic, for between 1973 and 1974 the fall in real wages was between 12.5% and 14%. For skilled workers there was another drop in wages in 1975, while there was no change for unskilled workers (who had experienced a greater drop over the preceding two years). As wages fell, unemployment rose and 'GNP fluctuated (see table 2.1).

By 1975 the inflation rate had fallen but there was a massive balance-of-payments deficit. There were purges in the bureaucracy and protests from church leaders and lawyers, armed rebellion and demonstration. From 1976 to 1980 there was a slow increase in real growth and the employment rate rose, yet the inflation rate exceeded the real growth rate and in the agricultural sector unemployment rose. In addition, the external debt rose (*ibid*). After 1981 inflation, wages and GDP rose and fell erratically. After Cory Aquino, by then Benigno's widow, came to power in 1986, there was a steady increase in GDP for a couple of years, but the value of agricultural crop production was still rising and falling.

Between 1970 and 1980 there was very little structural change in the Philippines economy. The undiversified nature of the economy was leading to stagnation as the level of activity in the traditional economic sectors of forestry, fishing and agriculture levelled (in agriculture, despite attempts at agrarian reform, activity fell). These sectors combined generated about one quarter of GNP, employed half of the labour force and earned more than 40% of export income (Commonwealth of Australia, 1983). While there was some fluctuation, GNP showed strong growth in the years 1972 to 1976. This continued with a rate of 6.7% until 1980 when GNP growth fell to 4.4%. GNP growth dropped again in 1981 to a rate of 3.8% (*ibid*).

Table 2.1 Basic indicators for the Philippines

Year	Real wages (% change)	Annual Inflation (%)	Crop production value (% change)	GDP growth (%) (1985 prices)
1982	10.2	10.2	2.4	1.8
1983	18.0	10.0	-2.1	1.2
1984	66.5	50.4	4.4	-5.8
1985	18.2	23.1	5.8	-3.8
1986	0.3	0.8	4.1	0.2
1987	8.2	3.8	-3.8	5.4
1988	14.1	8.8	2.3	6.3
1989	8.9	12.2	2.4	5.6
1990	8.4	14.1	-1.2	2.5

Source: Economic Policy Research Unit, 1994.

In August 1983 Benigno Aquino was assassinated at Manila airport as he disembarked from an aircraft. This triggered a political crisis which led to numerous demonstrations before the 1984 elections, which were won, in a sense, by both sides, with Marcos retaining control of the Legislature. Marcos called a snap election for February 1986, which led to Corazon Aquino, Benigno's widow, becoming President of the Philippines (Lande, 1986).[13]

The new president of the Philippines, Corazon Aquino, had come from a family that was by no means poor or underprivileged. She was born a daughter of the immensely wealthy Cojuangco family, whose assets included a 6000 hectare sugar plantation in Tarlac on the island of Luzon. However, the tenants and labourers of the family plantation, despite more promises of land reform, are still as poor as their peers all over the Philippines. For Aquino there was no charity at home (Goodno, 1991).

Table 2.2 Philippines inflation and external debt: 1972-82

Year	Annual inflation (%)	External debt ($U.S. m)
1972	16.4	2210
1973	16.2	2306
1974	34.5	2723
1975	6.9	3402
1976	9.7	5099
1977	9.4	6562
1978	7.5	8195
1979	17.5	9773
1980	18.2	12188
1981	13.1	14924
1982	10.2	17470

Source: Rodriguez, 1985.

In 1960 per-capita income in the Philippines had been one of the highest in the major market economies of the Western Pacific Region, which includes Korea, Taiwan, Indonesia, Malaysia, Thailand, Hong Kong and Singapore. By 1986 the Philippines had one of the lowest per-capita incomes in the region. In 1983 and 1984 GNP growth was negative, falling to a low of -7.08 in 1983 (see Table 2.3). The main cause of the fall in income and GNP growth was the drop in world prices for the country's major export commodities, sugar and coconut (see tables 2.4, 2.5 and 2.6) as well as rising external debt (see Table 2.2) and a high degree of dependence on fuel-oil imports (Commonwealth of Australia, 1986).

Table 2.3   Economic growth rates in the Philippines in the later Marcos years: GNP and GDP growth (%), and gross value-added growth (%) by sector

Year	GNP	GDP	Agricultural	Industrial	Service
1980	3.68	3.78	3.69	4.46	3.19
1981	1.55	2.90	3.13	2.15	3.47
1982	1.10	0.93	-2.10	0.67	3.20
1983	-7.08	-6.00	2.27	-10.22	-7.37
1984	-4.23	-4.39	3.32	-10.17	-4.65
1985	1.51	-1.08	3.74	-2.74	2.26

Source: Derived from Center for Research and Communication, 1988.

Table 2.4   Philippines' share of the world market in the later Marcos years (% of world value)

Year	Bananas	Coconut Oil	Sugar
1980	5.3	67.0	3.9
1981	5.4	64.5	3.5
1982	6.7	59.9	3.4
1983	5.0	72.2	2.7
1984	5.5	53.9	2.7
1985	4.6	41.8	1.8

Source: Derived from Center for Research and Communication, 1988.

It had been hoped that the new Aquino government would address the issues of a 'neglected, low productivity agricultural sector, a highly protected, inward looking manufacturing sector, and a lack of confidence in economic administration ...' (*ibid*, p. 80).

**Table 2.5 Real world prices for important agricultural commodities:
Constant 1990 prices**

Commodity	Year 1975	Year 1985	Year 1992	Price Unit
Rice	803.1	314.7	269.7	US $/mt
Sugar	1.0	0.13	0.19	US $/kg
Bananas	545.7	554.4	443.9	US $/mt
Cocoa	2.76	3.29	1.03	US $/kg
Coconut oil	870.4	860.1	541.9	US $/mt
Coffee	3.19	4.71	1.32	US $/kg
Copra	566.9	562.7	562.7	US $/mt

Source: Derived from World Resources Institute (1994).

By 1992 there was increased pressure on agricultural lands due to continuing population growth, the volatility of prices for cash crops sold on world markets, the failure of agrarian reform programs and the increasing poverty of farmers. The sustainability of land use practices continued to decrease and severe soil erosion in the uplands was getting worse, as evidenced by the effects of cyclone Thelma, which hit Leyte in November 1991.

Between 1984 and 1992 the total value of the Philippines major export crops fluctuated dramatically. The value of coconut exports fell steadily from 1984 to 1986, made a gradual recovery which peaked in 1989, and then dropped to a low in 1991 before showing some gains in 1992. Banana export value grew dramatically from 1987 and was still demonstrating good value in 1992. However, sugar export value has never returned to 1986 levels (see Table 2.7).

The Philippines is a price taker for sugar and bananas on world markets, and although it is the world's major producer of coconut products, it lost considerable influence when the U.S., its traditional market, substituted many home grown products such as safflower, sunflower and rapeseed oil for coconut oil in the early 1990s.

By 1989 world economic growth was contracting, with major industrial countries experiencing a slow down. However, although most countries experienced a decline, the Philippines was falling far behind some of its neighbours, such as Thailand (see Table 2.8), and the country was no longer considered to have a high-growth economy. Even in Southeast Asia the Philippines was being left behind in terms of world growth and development (International Monetary Fund, 1991).

Table 2.6 Change in world prices for selected agricultural commodities: Base year 1960 = 100

Year	B'na	Coco	Cnut	Cfee	Copr	Crn	Rce	Sugr
'64	119	86	95	127	94	100	110	184
'65	111	62	112	122	109	99	109	64
'66	108	88	104	111	89	94	131	58
'67	111	101	105	103	99	95	165	61
'68	107	122	128	102	112	104	162	61
'69	111	153	116	111	98	94	150	102
'70	116	115	127	149	109	101	115	117
'71	98	91	119	122	91	120	103	143
'72	113	109	75	139	68	127	118	232
'73	115	192	164	189	171	226	281	301
'74	129	265	320	200	320	305	435	945
'75	173	211	126	229	124	276	291	649
'76	180	347	134	408	133	260	204	368
'77	192	643	185	841	194	215	218	259
'78	201	578	219	396	227	232	295	248
'79	228	559	315	481	325	266	268	308
'80	265	442	215	568	219	289	348	913
'81	280	353	183	479	183	302	387	540
'82	261	295	149	396	152	252	234	268
'83	300	360	234	359	240	310	221	270
'84	259	407	370	394	343	n/a	202	166
'85	272	384	189	398	186	n/a	173	130

Source: derived from Center for Research and Communication, 1988.

Table 2.7 Export value of selected Philippines agricultural commodities ($US m)

Year	Coconut	Bananas	Sugar
1984	727	7	262
1985	459	7	256
1986	470	8	275
1987	566	121	71
1988	582	146	74
1989	772	220	179
1990	503	86	191
1991	447	173	136
1992	633	157	111

Source: Center for Research and Communication, 1994.

Table 2.8 GDP percentage growth rates for selected countries

Year	Philippines	Thailand	Major industrial nations	Australia
1973-83	5.0	6.6	3.7	1.8
1989	5.6	12.0	3.3	4.4
1990	2.1	5.6	2.6	1.6

Source: Derived from International Monetary Fund, 1991, and Commonwealth of Australia, 1986.

By 1992, when Aquino's term expired, insurgency and violence continued, poverty abounded, the gap between rich and poor was a chasm, and there had been no effective land reform. The New People's Army still had about 10,000 full-time members and an influence over about a quarter of the population. All over the Philippines there were movements of labourers, peasants and slum dwellers working together to improve their lot (Goodno, 1991).

Table 2.9   Summary of Philippines economic indicators in the later Marcos years

Year	Pop/n (m)	Labr Force (m)	GDP ($USb)	GDP/ capita ($US)	GDP grth (%)	CPI Chnge (%)	Blnce of trade ($USb)
'77	44.6	15.1	21.1	471	6.9	7.9	-0.8
'78	45.9	17.4	24.3	530	5.9	7.6	-1.2
'79	47.2	18.4	29.6	627	5.8	18.8	-1.6
'80	48.5	18.5	35.7	736	5.2	17.6	-2.1
'81	49.5	19.0	38.1	776	3.7	13.1	-2.3
'82	50.7	20.0	39.4	777	2.8	10.2	-3.4
'83	52.1	20.5	34.1	655	1.3	10.0	-2.8
'84	53.4	na	32.2	603	-5.3	50.4	-1.3

Source: Commonwealth of Australia (1981 to 1986).

When Aquino's term expired she did not stand for re-election. There were many candidates for the presidency in the May 1992 elections, including Imelda Marcos, Ferdinand's widow, who had returned home. The main contenders were Fidel Ramos, who had Aquino's endorsement; Eduardo Cojuangco, Aquino's first cousin, coconut industry monopolist and a former Marcos crony; and Miriam Defensor-Santiago, a lawyer and minister of Aquino's government, who ran on an anti-corruption ticket. Among cries of fraud, vote-buying, corruption and vote-rigging, Ramos was declared the election winner.

The Post-Aquino Years

Following Aquino, Fidel Ramos succeeded to the Presidency of the Philippines, with Joseph Estrada as his vice president. Having gained only 23.6 % of the vote, Ramos may have felt some apprehension at his inauguration on 30 June 1992. Ramos' stated principles for the country included an agenda for political and civil stability, deregulation of the economy and encouragement of foreign investment by dismantling monopolies, and the control of corruption and 'criminality' (Pinches and Brown, forthcoming).

Subsequent to the presidential inauguration, various shifting coalitions were formed to give Lakas – NUCD leadership in Congress in July 1992. While Ramos' tried to steer the country's economy towards deregulation, he was also maintaining talks with various insurgent and rebel groups to establish ceasefire agreements as part of his civil and political stability policy. These deregulation and stability goals were to set the scene as priorities for the duration of his term in office. Along with simmering military rebellion and communist insurgency, Moslem secessionism in the archipelago's south had posed a challenge to Philippines' governments for decades (McKenna, 1998).

By November 1992 the United States of America was completing its withdrawal from the Subic Bay Naval Base, prompting Ramos to request a review of the Philippines-U.S. Mutual Defense Treaty, which was ratified in 1951, for fear of what might happen if the Philippines were subject to external attack. However Ramos' greatest problem continued to be internal unrest and domestic violence, with the Philippines Left organising a protest over the increase in petrol prices at the start of 1994. This increase had resulted from a revenue raising government levy on petrol the previous year (Pinches and Brown, forthcoming). A number of bomb attacks on leading petroleum companies, and the threat of a general strike, forced Ramos the remove the petrol levy in February 1994.

Meanwhile, talks with the secessionist Moro National Liberation Front (MNLF) had progressed through 1993 with the MNLF demanding an independent Moslem state. However, much as usual, Philippines' life continued against a backdrop of violence with ongoing attacks and retaliations being exchanged between Christian and Moslem extremists (*ibid*).

The Marcos legacy continued to be costly to the government, and in February 1994 thousands of victims tortured under the Marcos regime were awarded massive total punitive damages. Later that year, the Ramos government began an offensive against Abu Sayyaf, a fundamentalist Moslem group that had been responsible for acts of kidnapping and terrorism in the southern islands. This government offensive was to prove ineffective as Abu Sayyaf activity in later years was to reveal, with kidnappings and assassinations attracting more international attention during the terms of two subsequent presidents.

In September 2001, following the terrorist destruction of the World Trade Center in New York, world notice was drawn to the Philippines. When the USA identified Osama bin Laden as the chief suspect in the New York disaster, the media reported a link between Abu Sayyaf and bin Laden, claiming that the Philippines group had received funding from bin Laden's organisation (Lyall, 2001).

From 1994-97 the government continued to negotiate with the MNLF and the MILF (Moro Islamic Liberation Front) over an autonomous Islamic Region on the southern island of Mindanao. Abu Sayyaf, opposed to the peace talks between the Moro separatists and the government, used violence as a means to disrupt the negotiation process.

With a general election approaching, Ramos supporters campaigned for constitutional changes that would have permitted President Ramos eligibility to stand for a second term of office. However this campaign failed, and on 11 May 1998, following a relatively quiet election, Joseph Estrada, former film actor and darling of the 'masses' was voted President of the Philippines with nearly 40 % of the vote.

Estrada's Cabinet and his strange consortium of political allies were drawn from wealthy, mainly ethnic Chinese, businessmen; Marcos cronies; former left-wing activists of various loyalties, and old friends. Given his 'pro-poor' election policies and his popular electoral appeal, Estrada formed some interesting political relationships. Not the least of these was with Eduardo Cojuanco, Marcos crony, Aquino cousin, 'coconut king', once and future Chairman of the San Miguel Corporation, and the subject of government litigation.

Having restored his friend Cojuanco to the position of Chairman and CEO of San Miguel in July 1998, Estrada set about disbanding the Presidential Commission on Good Government, formed to bring Marcos cronies to justice under Aquino (Pinches and Brown, forthcoming). This act was to foreshadow the tenor of Estrada's presidency and the manner of his downfall.

Estrada's main 'pro-poor' election platform had been to promise a programme of poverty alleviation, agricultural development and food security. This programme sat uneasily with his free-market, economic rationalist policy agenda and soon led Estrada into difficulties.

Meanwhile, as under previous governments, there was ongoing pressure from Moslem groups in western Mindanao to secede. Estrada sought to increase the number of provinces subject to Moslem autonomy by proposing to amend the Organic Act of the Autonomous Region in Muslim Mindanao (ARMM), but continuing problems in the area negated this intention (*ibid*). Sporadic violence and ceasefires between the MILF and government forces alternated through 1997 and 1998, but peace negotiations continued.

While Estrada demonstrated better leadership than expected in some areas of his presidency, there were growing concerns about cronyism and corruption, and about the extension of deregulatory policies to include the removal of foreign ownership limits on land and public utilities.

In addition, the poor seemed to be no better off and, despite ongoing peace talks, hostilities between government armed forces and the MILF escalated early in 1999, with the army attacking an MILF training camp. This apparent act of aggression killed and wounded numerous people and displaced thousands. Conflict between trade unions and the government over broken wage promises, on the one hand, and between business and the government over slow privatisation and liberalisation reforms, on the other, further tarnished Estrada's Presidential image through 1999 and 2000.

Hostilities in the Moslem south continued to increase and by July 2000 peace seemed unattainable. Abu Sayyaff augmented its attraction for global attention by kidnapping foreigners, which also netted the group millions of dollars in ransom payments.

In October 2000 Joseph Estrada was accused of taking massive bribes, and impeachment proceedings against him began. Earlier that month, Estrada's Vice-President Gloria Macapagal Arroyo had resigned her Cabinet position, but not her Vice-Presidency. In November Estrada was impeached for bribery, corruption, betrayal of public trust and culpable Violation of the Constitution (Pinches and Brown, forthcoming).

Gloria Arroyo succeeded Joseph Estrada as President of the Philippines, stepping into his place as he tumbled from power.

Agrarian Reform

In the Philippines, agrarian reform has been on the social and political agenda for many years. Attempts to narrow the gap between the landed and the landless have given rise to various political movements over the course of this century (see earlier sections of this chapter), the best known of these in contemporary times being the New People's Army. Successive governments have promised to reform land tenure and ownership, but somehow no policy has ever been implemented that has had a lasting effect on the socio-economic status of the majority of farmers. The importance of land tenure cannot be overemphasised in the socio-economic context. Christodoulou (1990) states:

> Land is at the heart of agrarian conflict and reform and is the pivot of power. More precisely, the problem is one of human or social relations in respect to control and use of land and access to the accruing benefits. The land system is a network of these relations in institutionalised form. Land becomes the pivot of power because people depend on it for their vital

needs: the greater the dependence, the more strategic becomes control of land and the more power it confers. Monopoly control of land may lead to monopoly control of power over people (p. 1).

The present system of land ownership and land tenure has its origins in Spanish colonial administration and the patronage of the Catholic church, as discussed earlier in this chapter. In the Philippines, and indeed at our study sites specifically, non-land owning farmers may be tenants, paying a flat-rate rent regardless of the variations in production, or share farmers who pay a proportion of their harvest (up to 70%) to the land owner. Some consider the tenancy system to be more just than the share system because the tiller of the land benefits from the increased effort and good fortune of a plentiful harvest (although the converse also applies), rather than having to pay a proportion of any increase over to the landlord.

There are advantages and drawbacks to both the tenant and sharefarmer systems, although in either case the gains are weighted in favour of the landowner. In the 1950s and 1960s in the Philippines attempts were made to legislate in favour of converting sharecropping to tenancy on the premise that this was a progressive step (*ibid*). However, these attempts at agrarian reform did little to redress the imbalance of the privilige of the landlord and the poverty of the farmer, or give farmers greater control over the lands they tilled.

According to the 1980 census, 63% of the population of the Philippines lived in rural areas and the majority of these people were poor (Arce and Abad, 1986). Arce and Abad (*ibid*) state that:

> The limited effectiveness of Operation Land Transfer – the latest in a series of generally emasculated agrarian reform programs – is the most recent demonstration of the weakness of the disadvantaged in the rural areas. To be fair, credit should be given to the Ministry of Agrarian Reform, by August 1982 certificates of land transfer had been issued to 430,273 farmer beneficiaries, a total that exceeds the target set by the Ministry for the whole program. But the same ministry report cites an even higher number of farmers (609,042) as being eligible only for leasehold arrangements, not for eventual ownership of the land they till. Further, these two groups of tenant farmers are still only a portion of those who derive their living from the land but do not own a piece of it. the number of these in 1975 was estimated at 3.5 million (p. 61).

When Marcos had come to power in 1965 he had promised many reforms, including land reform. He soon moved to implement the existing 1963 Land Reform Act in Pampanga Province as a result of Huk resurgence. However, the sincerity of his land reforming acts may have been questionable – for example, his four-year land distribution budget was equal to his 1967 annual military budget (Noble, 1986).

Violence within the nation continued to rise. By 1972, the NPA was very active in Central Luzon and it continued to spread across the country, particularly to areas where farmers had little control over their lands and were subject to tenancy abuse. Post 1975, the Marcos regime described agrarian reform as 'the cornerstone of its development efforts' and was eager to improve the productivity of corn and rice. Noble (*ibid*, pp. 90-91) comments:

> Essentially, the land reform program was directed at all rice and corn holdings above seven hectares in size, and was intended to provide tenants on these holdings (with) family sized farms (3 hectares if the land was irrigated, 5 if it was not). Tenants were to be distributed land transfer certificates, land prices were to be fixed at 2.5 times the value of the average gross harvest, and payments were to be amortised over a 15-year period ...
>
> Meanwhile the former tenants would be assisted through bank loans (commercial banks were required to make a fixed %age of their loans for rural purposes); through their membership in barrio associations called Samahang Nayons, which they were required to join before being given land transfer certificates.
>
> The food production drive was carried out primarily through a program called Masagana 99, which was intended to triple rice yields/acre. The program provided cheap credit for a seed/fertiliser/pesticide input mix specified by the government.

Obviously this particular reform program was very limited as it did not assist tenant farmers on farms smaller than 7 hectares, nor did it help the landless labourers, or producers of anything but rice and corn. And even within its narrow scope it was not very successful. In other reforms, attempts to improve the price of sugar by witholding stocks were also disastrous, leaving the government with massive stockpiles that it was unable to store.

By 1980, with no obviously successful attempts at agrarian reform behind it, the Marcos government appeared to be concentrating on other issues, and by then it was estimated that 70% of the population was malnourished (*ibid*).

When President Aquino came to power she made promises to introduce agrarian reform, and much of her support came from the rural base. However her promises were never implemented despite the Ministry of Agrarian Reform and the Ministry of Agriculture consulting farmers' organisations. The problem of inequitable land distribution was recognised by the government and by the World Bank (see Goodno, 1991, p. 270) but the government continued to prevaricate about the depth of its commitment to any reformist program.

The matter came to a head in January 1987 when 13 pro-reform demonstrators were murdered outside the presidential palace. A cabinet committee was appointed to address the issue in the face of vehement opposition from landholders and Aquino gave the power of decision-making on matters such as timetables for land transfers to the landholder-dominated Congress. A Comprehensive Agrarian Reform Act was signed in 1988. Though it had an impressive title, the Act did little to address the imbalance in land ownership distribution (*ibid*). Goodno cites the Congress for a People's Agrarian Reform (CPAR- a broad based pressure group) objections to the new Act as follows:

> 1. The retention limit of 5 hectares for the landowner automatically exempts 51% of the total private agricultural land area. In addition, qualified heirs of landowners retain 3 hectares each, whether actual tillers or not. Assuming an average of two legal heirs per family, at least 75% of all private agricultural lands will not be covered. Moreover the scope of the CPAR is reduced even further by the 10 year reprieve given to commercial farms or lands devoted to commercial livestock, poultry and swine raising, aqua culture and orchards, among others.
> 2. Landlord corporations that merely distribute part of their stocks to tillers are not obliged to redistribute their lands to tillers/farm workers. Under this scheme, farmer workers will be getting only a minimal share in the corporation and will have no significant participation in its decision making.
> 3. Final determination of compensation rests solely wiwth the landowners and the Department of Agrarian Reform and the Land bank of the Philippines. The farmers have no say.
> 4. All public agricultural land used by multinational corporations (MNCs) shall be distributed to the workers within three years, beginning the year the law shall take effect. Succeeding provisions, however, allow MNCs to continue leasing the land with the approval of the government or the beneficiaries to whom the lands may be transferred. Furthermore, section 72 allows the lease, management, grower or service contracts in private lands to continue even if these lands have already been transferred to their beneficiaries. These are the loopholes through which foreign and domestic corporate interests can retain their control over vast land holdings (p. 272).

Meanwhile, attempts at reform in the regions failed to make any great impact. The Balanced Agro-Industrial Development Strategy (BAIDS), the *Kilusang Sarilang Sikap* (KSS) and the *Kilusang Kabunayan Kaunlaran*

(KKK) programs, developed to address rural poverty, loss of economic status and problems in agricultural production, seem to have had little permanent effect in the Eastern Visayas Region (Regional Development Council, 1985).

The major reform groups in the Philippines believe that equity and social justice should be the goals of agrarian reform and that farm workers should be given land free. The Congress for a People's Agrarian Reform (CPAR), the Peasant Movement of the Philippines (KMP), and the Communist Party of the Philippines/National Democratic Front (CPP/NDF) believed that the following steps should have been followed as an interim measure (Goodno, 1991):

- rent reduction
- the elimination of usury, and
- the transfer of the burden of paying for farm inputs to the landowner.

However, there has been little progress in agrarian reform over the past four decades, and Estrada's broken promises for poverty alleviation, food security and agricultural development did nothing to increase the confidence of the rural poor. During Estrada's term, Horacio Morales was appointed as the Secretary for Agrarian Reform. Morales made a commitment to implement fully the CARP land reform programme to redistribute 1.3 million hectares of land, that had started more than ten years earlier. Little headway was made, and the CPP/NDF leadership condemned Estrada's government.

Overall, the government's fulfilment of its 'pro-poor' policies resulted in little more than the establishment of state bodies with impressive names. The main aim of the 'pro-poor' platform was to ensure food security for the impoverished Filipinos through an agrarian reform programme that centred on 'agricultural modernisation' through the passage of the Agricultural and Fisheries Modernisation Act. This Act provided budget allocations that were meant to provide sufficient funding for rural credit, enhanced access to irrigation and improved rural infrastructure. Neither these commitments, nor the promised land redistributions were fulfilled. There were political consequences for the Estrada government as a result of the failure to implement pro-poor and agrarian reform programmes. The communist movement, while dispersed and minor in the late nineties, was able to recruit many disillusioned voters, and the rural poor of Mindanao are the major constituency of the Moro Islamic Liberation Front (Pinches and Brown, forthcoming).

The repeated failure of successive Philippines' governments to fulfil promises for Agrarian reform is likely to have long-term repercussions for a nation that has a history of corruption, insurgency, poverty and rebellion.

Notes

1.   While there is not a direct parallel with this system today, barangay life is by no means homogenous, but highly stratified, and there is probably still a fairly strong kinship role in the determination of social standing.

2.   Indeed in the central Philippines there is a rich archaeology of Chinese ceramics (Solheim, 1974).

3.   The CAA attributes the loss of this heritage to the Spaniards (and it is likely, given the heavy influence of the church in education in the regions, that it was specifically the Friars who were responsible for this loss).

4.   Although Magellan was originally Portuguese. His patron had been the King of Portugal, but after a disagreement in 1505 or thereabouts he changed his allegiance to Spain, and made a pledge to the King of Spain to search for the Spice Islands within the limits of the Spanish Empire. It was this quest that led him to the Philippines.

5.   Today's slash-and-burn agriculture, occurring in the mountain forests, is by no means as friendly to the ecosystem as this early version.

6.   The system of settlement and land dispersal used by the Spaniards in the Philippines was not very different from that which they used in South America. In El Salvador, which the Spaniards began conquering in 1524, the system of encomienda was introduced and then superseded by the hacienda plantation approach. The church also had a role in this settlement and land use pattern. However, in El Salvador the immediate impact on the country's inhabitants, who were less dispersed than those of the Philippines, appears to have been more severe (Pearce, 1986).

7.   Eventually the Spaniards and the Americans agreed to a secret treaty whereby the islands would be surrendered to the USA in a mock battle. The Philippines independence movement was not party to this agreement. The Spaniards and the Americans now had a partnership that excluded the Filipinos (McCoy and de Jesus, 1982). The revolutionary leaders continued to press for independence when it became apparent that the Americans had no intention of leaving the islands, and a war of resistance broke out in February 1899. In March 1901 Aguinaldo was captured by the Americans and finally swore allegiance to them. However, resistance to the USA continued until 1903 (Worcester, 1921). During the period of confrontation between the Spaniards, the Filipinos and the Americans, the Moslems of Mindanao and Sulu (the Moros) had remained neutral, having no allegiance with the Christian Filipinos, and in 1903 the Americans established a Moro Province. However, the occupation was viewed with dread by the Moros and they continued to resist until 1914 when the major Moslem groups were put down by the American military rulers of the province.

8.   The American influence on the Philippines is still strong. One of the main languages spoken in the country is English (Spanish was never widely spoken among the inhabitants); there were ties and treaties with subsequent governments up to the present; the education system was strongly influenced by the Americans;

American aid was forthcoming; and American missionaries are spread across the islands.

9. In 1948 the Huk was outlawed and in the same year President Roxas died of a heart attack and was succeeded by President Quirino. In 1951 the U.S.-Philippine Mutual Defence Treaty was signed and in 1953 Ramon Magsaysay, a former guerilla, beat Quirino in the presidential election. Magsaysay attempted to improve the land tenure system but achieved very little in the long term. He was killed in an aeroplane crash in 1957 and Carlos Garcia won the presidency in November that year (Davis, 1987; Goodno, 1991). In 1961 Garcia was defeated in the elections, on the issues of graft and corruption, by Diosdado Macapagal, who was not from the ruling elite. Macapagal, who understood, and was concerned with, the historical plight of the peasant farmers, passed a law for the Amended Land Reform Code in 1963. Again, little was achieved, although rebellion and insurgency had continued. The power of the Huk died in 1964 as a result of promises of social and economic reform. Ferdinand Marcos defeated Macapagal in the Presidential election of 1965 (Davis, 1987).

10. Ferdinand Marcos grew up in Ilocos Norte, on the South China Sea. He was the son of a teacher turned lawyer and failed politician. In 1938 Ferdinand Marcos was arrested on suspicion of having murderered one of his father's more successful political rivals, Julio Nalundasan, in 1935. In 1940 Marcos was found guilty of the killing and was sentenced to 17 years in prison; in the same year Marcos passed his bar examinations. He presented his own appeal to the Supreme Court and had his sentence quashed. However, many Filipinos still considered him to be guilty. Marcos entered politics in 1949 (Goodno, 1991). Thus began the career of Ferdinand Marcos, who was to become one of the most notorious dictators of the 20th century. Stories and rumours about Ferdinand Marcos's origins and early years abound. One of the persistent rumours was that he was illegitimate, and not a Marcos son. Seagrave (1988) asserts that his biological father was a Chinese magistrate and 'leading member of one of the six richest and most powerful clans in the islands, a billionaire clan involved in the daily financial, commercial and political transactions that are the lifeblood of the islands' (p. 5).

11. Seagrove (1988) sees allegations about the identity of Marcos's true father as being significant to the style of his government. He says:
'Once the stature of Ferdinand's father was confirmed a number of other riddles were solved: How a young Ferdinand eluded a murder conviction in his schooldays. How a place came to be waiting for him in a brotherhood referred to as the Ilocano Mafia, whose pre-war enterprises were said to include smuggling, extortion, black marketeering and murder for hire. And how, after the war, Ferdinand became a young congressman with extraordinary connections in the Chinese financial world, using his position in congress to extort large sums from Chinese businessmen. The leverage of his father's clan enabled Ferdinand to ally himself secretly with agents of the Chiang regime, with Japanese underworld syndicates, and with some big-time American operators' (p. 6).

12. Years later, Enrile confessed that the ambush had been staged by government men (Goodno, 1991).

13. The 1986 election result was not a simple matter. The Marcos controlled National Assembly declared Marcos to be the winner, but other projections and estimates (including one leaked from the CIA) gave Cory Aquino up to 58 % of the vote. Aquino rallied her supporters and mass anti-Marcos demonstrations were held. These heralded the start of a 'people power' campaign of civil disobedience, and a

boycott of all banks, media and corporations controlled by Marcos cronies. Throughout the latter part of February plots and counter plots prevailed as a result of action by various opponents of the Marcos government. On 22 February there was military movement as news of a dissident plot reached the Marcos camp. President Reagan of the USA eventually called for the resignation of his friend Marcos. On Tuesday 25 February 1986 the Marcos era ended and Cory Aquino was sworn in as President of the Philippines. The Marcos family fled to Honolulu in the USA, allegedly taking with them $40 billion worth of bullion and other assets (Goodno, 1991; Seagrave, 1988).

Chapter 3

The Philippines and Leyte

Introduction

This chapter provides the background for the socio-economic analysis of factors influencing decision making covered later in the book. In it the risks associated with weather conditions, harvest failure, price volatility, land degradation and food availability are discussed. The conditions in Leyte make it appropriate for a case study on sustainable land use and farmer decision making because they reflect the socio-economic and physical situation for farmers in many less developed countries.

In the Philippines, soil erosion is increasing on agricultural lands, particularly in the unstable uplands, where the rate of adoption of soil conservation methods is low. In addition, the effects of tropical cyclones is becoming worse as a direct result of land degradation. Population growth is accelerating and the marginal uplands are increasingly cultivated to feed the growing population. Poverty is severe and income distribution is inequitable. In addition, the agricultural outlook is not improving, with losses in the world market share of the Philippines' major cash crop, coconut products, while agricultural commodity prices continue to be volatile.

Description of the Philippines

The Philippines lies in the Pacific Ocean and is made up of approximately 7100 islands and islets. The archipelago is spread over 1.29 million km^2 of ocean and occupies 299,400 km^2 of land. Ninety six % of the land area is occupied by 11 islands, and more than half of the population of 70 million lives on the two largest islands, Luzon and Mindanao (Davis, 1987; Goodno, 1991).

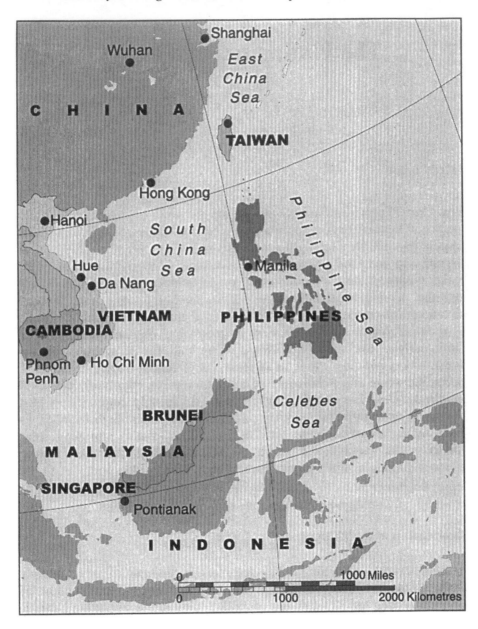

Figure 3.1 Location of the Philippines in South-East Asia

The main island, on which the capital, Quezon City, is situated, is Luzon, the most northerly island of the group. The distance from the northern tip of Luzon to the southernmost point on the Sulu archipelago is about 1625 km. Due south lies Sulawesi, southwest is Kalimantan and west across the South China sea is Malaysia. About 1295 km north and north-west of Luzon lies China (see Figure 3.1) (Fullard, 1979).

The Philippines lies on the margin of South-east Asia, with access to the islands of the Pacific. Being an oceanic nation, the Philippines has been visited by seafaring peoples for thousands of years, many of whom migrated there from their homelands (see Chapter 2).

The islands fall into arbitrary divisions constituting Luzon and Mindoro in the north, Palawan in the west, Mindanao and the Sulu archipelago in the south and the Visayan islands at the centre. Leyte on which this study focuses, is part of the eastern Visayas (see Figure 3.2).

There are many physical similarities between the Philippine islands and the mainland countries of South-east Asia, including some vegetation, soil characteristics, climate and geomorphology. However, even among the islands there is a good deal of variation, and on the island arcs of the Pacific where the ocean has great influence, the physiography and topography is dissimilar to that of mainland Asia (Wernstedt and Spencer, 1967).

Rainfall varies widely over the islands and vegetative cover ranges from grassland to tropical rainforests as a result of variation in topography, rainfall and soil type. The country has some extensive river systems such as the Pulangi and the Agusan in Mindanao, the Cagayan in northern Luzon and the Pampanga, which flows into Manila Bay. There are also a number of freshwater lakes (Edgerton, 1984). The river systems are used for irrigation in wet rice cultivation, which is spread across the islands.

Many of the islands' rainforests have been dislocated by logging activities and fruit plantations (Saastamoinen, 1996; Uitamo, 1996). Other resources of the archipelago include minerals such as copper, gold, zinc, nickel, iron, manganese and chromite, though these occur mainly in Mindanao, Surigao, nothern Luzon and Cebu.

The Philippines is densely populated and the total population of the islands has experienced a more than one hundredfold multiplication since Spanish settlement. The population of metro Manila is currently about 6 million.

The Visayas

The Visayan group of islands is placed near the centre of the Philippines Archipelago. It includes the larger islands of Samar, Leyte, Negros, Panay, Cebu, Bohol and Masbate as well as several small islands. The name Visayas derives from the old tribal name of the islands' main residents, the Visayans, whose name, in turn, is believed to derive from the Javanese Sri Vijayan Empire, which was transcendent in the seventh century and prominent for 600 years. This empire held part of the Philippines under political control for some time.

There are four main related dialects spoken by the Visayan people. Of these two are spoken on Leyte: Cebuano in the south and west, and Waray-Waray in the north and east (Wernstedt and Spencer, 1967).

Leyte is in the Eastern Visayas, which comprise five provinces: Eastern, Western and Northern Samar, Leyte and Southern Leyte. The main islands of the Eastern Visayas, Leyte and Samar, are separated by the San Juanico Strait. To the north of Samar is the tip of the Bicol Peninsula, while to the east of both islands lies the Pacific ocean. On the Western sides of Leyte and Samar are the Visayas and Camotes (sweet potato) Seas and Maquada Bay, and to the south of Leyte is Surigao (see Figure 3.3).

Leyte is the largest of the provinces in the Eastern Visayas region, with an area of 6268 km^2, and Southern Leyte is the smallest, measuring 1735 km^2. The total land area of the region is 21,563 km^2 (Regional Development Council, 1985). There are clusters of small islands off the coasts of Samar and Leyte, but the main islands comprise 93 % of the region's land area.

Samar is the third-largest island in the Philippines Archipelago, while Leyte is the eighth-largest. Both of the islands are rugged and mountainous. Leyte has three main mountain systems, all of which have a volcanic past. In the north western corner of the island lies the Northwestern Highlands, below which Villaba (one of the study sites), among other barangays, sits on the coast. The Central Cordillera, which rises to over 1400 metres at its highest point, runs down the centre of the island almost parallel to the western coast. The Northeastern Highlands, which are shorter and lower than the other ranges, look out to Samar over the San Juanico Strait.

The structure of Samar is somewhat different from that of Leyte, having a broad central plateau with steep valleys cut by the island's many streams. The highlands on both islands are sparsely settled and densely forested, with 75 % of the overall region forest-covered.

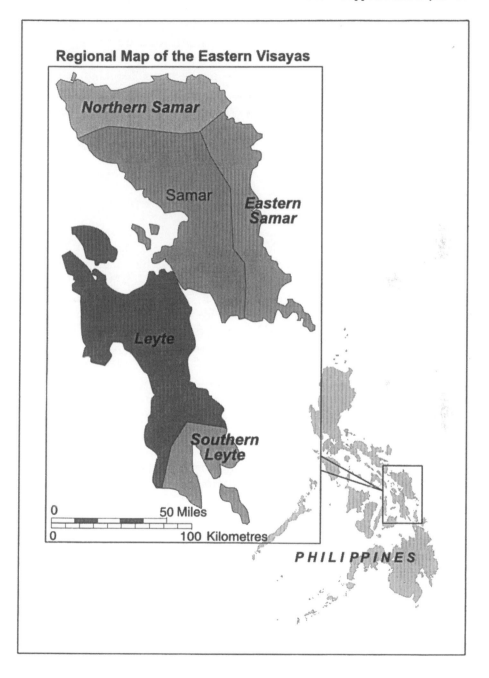

Figure 3.2 Regional map of the Eastern Visayas

However, as population pressure continues, and flat agricultural land becomes more scarce, the farmers of the coastal fringes are pushing further into the highlands to cultivate their crops.

Generally the Eastern Visayas region is tempestuous, blocking as it does the Visayan Sea from the open Pacific. There are about five cyclones in every five-year cycle and these may be accompanied by winds over 120 km/hour. The seasons are not sharply defined although the wettest months are usually from November to January, with the driest months running from April to August (*ibid*).

Description of Leyte

As discussed above, the Eastern Visayas islands are very hilly and most are traversed by mountain ranges. Leyte, with its three mountain ranges, has only one extensive lowland area, the Leyte Valley, which spreads between the Northeastern Highlands and the Central Cordillera. The lowlands are already cultivated; the only place left for the growing population to establish additional crops is on the steep upland slopes. The Leyte Valley lies west of the city of Tacloban, the largest city on the island, and runs north-west to Carigara Bay and south-east to Leyte Gulf. Most other lowlands on Leyte run in thin strips aroud the coast, or border the rivers in the form of small alluvial plains or deltas. Leyte Valley is an alluvial plain, recent in geological time, which was once a shallow ocean arm joining Carigara Bay and Leyte Gulf. The plain is traversed by rivers which drain into the Bay or the Gulf.

The other main lowland in Leyte is the Ormoc Valley, which runs north from the city of Ormoc. The valley is drained by the Pagsangahan River, south into Ormoc Bay.

Much of the climate on Leyte is differentiated from that of other Philippines' islands (except Samar, which is very similar to Leyte) by the sheltering effect of its mountains on the one hand, and by its proximity to the open Pacific Ocean, which exposes it to north-eastern air masses and cyclones, on the other hand. In the winter period (as far as there can be such a season) when the sun is low, there is very heavy rainfall in the north and east of Leyte. Because of the mountain ranges between the west and south of Leyte and the Pacific Ocean, these areas experience a relatively low rainfall in the winter season. The whole island receives a high rainfall when there is south-western air movement in June, July and August, although this is lower in the east, which is more sheltered at this time (Wernstedt and Spencer, 1967).

Figure 3.3 Map of the Eastern Visayas

Generally speaking, the climate of Leyte is hot, wet and humid. There is no season of drought (although droughts do occur, for example in 1987) and rainfall is high throughout the year. This means that Leyte has been able to support an agricultural base that is not strongly seasonal. The steady rainfall brings some disadvantages in that where the soils are impermeable, such as in the Leyte Valley, there is a propensity to flood or swamp. In addition, because of population pressure and food shortages, many farmers are moving up from the coastal lowlands onto the unstable hillsides, which have a tendency, in times of cyclones and heavy rainfall, to slide away, sometimes with devastating effect. One of the dubious distictions attributable to Leyte is the frequency and destructive nature of its cyclones. These cyclones may occur at any time of the year and result in floods, landslides and sometimes tidal waves, which cause serious damage to buildings, crops and land, and loss of life, sometimes on a large scale (as happened around Ormoc in November 1991). With respect to land resources, they probably also cause severe long-term damage to the ability of survivors to maintain themselves, since the area of land available for subsistence agriculture is so small.

Leyte still has extensive forest resources, although these are being encroached upon by forestry and agricultural operations (Saastamoinen, 1996). The centres of the sawmill industry are in the Central Cordillera and at the south of Leyte Valley. The major remaining forests are in the Central Cordillera and in south-east Leyte.

Leyte does not have a significant base of mineral deposits, although at Villaba (one of the study sites) there has been exploitation of sandstone and limestone impregnated with crude oil, which is usually used for road building around the island. There is also a manganese deposit near Baybay which has been used from time to time (*ibid*; National Economic and Development Agency, 1985).

Natural Disasters

The Philippines lies in a part of the world that is prone to natural disasters. Apart from spectacular disturbances such as the recent eruption of the volcano Mount Pinatubo, the Philippines is subject to about 20 typhoons a year. A typhoon, also known as a tropical cyclone or hurricane, forms over a warm ocean. The tranquil centre (eye) of the storm contains warm air which contributes to low pressure at the surface. Warm, moist air circulates around the eye and the typhoon can have a diameter as wide as 480 km.

Very high winds with a velocity of up to 320 km/h, and heavy rains, accompany the typhoon.

Every year hundreds of people in the Philippines die as a result of typhoons. However, while there is some expectation of wind and water damage and flooding across the islands, the impact of typhoons appears to be getting worse, as they are now accompanied by very destructive landslides.

On 4 November 1991 tropical cyclone Thelma, known locally as typhoon *Uring*, hit the Visayas and in three hours caused major damage to the Leyte city of Ormoc. A week earlier, tropical cyclone Ruth had hit the Cordillera provinces of the Philippines, causing landslides and mudflows in Baguio City and areas of Central Luzon already damaged by the eruption of Mount Pinatubo.

The later cyclone, Thelma, ultimately killed about 8000 people in and around the city of Ormoc, most of them swept away in floods and landslides, or crushed to death in their homes under a mass of mud. This was the highest death toll from a single natural disaster in the Philippines since a tidal wave along the Moro Gulf of Mindinao killed 8000 people in 1977.

Cyclones and Deforestation

Apart from causing massive socio-economic costs in terms of loss of life and physical destruction, Cyclone Thelma left controversy in its wake. For the month following the disaster arguments raged in the Philippines' press. While the central reason for this debate was the apportionment of blame for the scale of the tragedy, it became clear from meteorological reports and the extent and nature of the damage that the force of the storm alone would not have been expected to have left such a massive impact. Very soon after the storm, questions were raised about the synergistic effects of landuse in the mountains around Ormoc.

According to a bulletin by the Citizens Disaster Rehabilitation Center (CDRC) in Quezon City (9 November 1991) the extent of the damage was due to 'the logging of the dipterocarp forests'. The CDRC also stated that ... 'The government meteorological agency PAGASA ... had been hard put in explaining the disaster that hit the Visayas', and that ... 'PAGASA [had previously] claimed that the mountainous terrain of Leyte protects it from storm surges. National dailies would later report that government meteorologists are belatedly 'speculating that the coastal area (of Ormoc City) experienced a storm surge'.

While the government of then President Cory Aquino placed all the blame on illegal logging operations, contemporary media reports blamed the massive impact of the storm on 'virgin forests ... slowly being destroyed by poor Filipino peasants forced to practice slash-and-burn farming' (*Philippines Daily Globe*, 8 November 1991) and 'rapid deforestation' (*Daily Inquirer*, 10 November 1991). The *Daily Globe* editorial 'Still to blame' (13 November 1991) apportioned responsibility to legal and illegal loggers and to the government, while the *Manila Chronicle* editorial 'Gov't must react with firm policy, not panic' (15 November 1991) also blamed legal and illegal logging activities but went further in its discussion of the problem, stating:

> ... it has now become quite clear that the actions of government agencies appear to contradict policies regarding the restoration of the integrity of our natural environment. For instance, while it has been a long-standing policy to substantially restrict logging activities, the Department of Environment and Natural Resources has continued to issue permits. As a consequence, our country's forest cover has continued to recede from 17 million hectares in 1934 to a distressing 6.1 million hectares in 1990.

In an editorial entitled 'Fingering fingerlings' (13 November 1991) the *Philippines Daily Inquirer* reported the governor of Leyte, Leopoldo Petilla, as saying that the denudation of forests was the main cause of the terrible impact of the cyclone. The paper stated that Mr Petilla:

> said the rape of Leyte's forests was perpetrated by businessmen working in connivance with the government and military officials. Ormoc City Mayor Maria Victoria L. Locsin also said the same thing. And so did Cebu governer Emilio Osmena, who was among the first to go and bring aid to the devastated city ...
> The Department of Environment and Natural Resources said the watershed area in Leyte had become 'highly erodible' and traced this to the conversion of forest lands into agricultural uses starting in 1952.

While it is obvious from reading the media reports that the argument about who is to blame for the scale of the disaster is highly emotional, it is clear that all parties agree that the severe landslides and flooding stem from inappropriate land-use in the hills. The primary cause was loss of forest cover and the cultivation of denuded land for agriculture, leading to destabilisation of the hillsides and decreased water uptake by tree roots during storms. This in turn resulted in increased runoff in the water catchments and flooding of the rivers.

According to media reports the storm was not severe in the context of the norm for cyclones. The physical sequence of events, as reported in the *Manila Times* (7 November 1991) was as follows:

> Uring, whose peak strength of only 60 kilometers per hour was not even enough to qualify it as a typhoon, struck Central Visayas unexpectedly. It had changed course, hitting land at Samar, instead of Albay. The storm released little rain as it crossed Samar, but unleashed its full fury over Leyte, the weather bureau said.
>
> ... the rains triggered landslides along mountains largely stripped of forest cover after years of heavy logging. Floods heavy with mud and debris then tumbled down the slopes, swelling four major rivers and the coastal waters off Ormoc. In Ormoc a dam burst, washing away an entire village.
>
> ... overflow from two rivers – Marabong and Daguitan – merged and cascaded into the landlocked area.

Ironically, following the cyclone the major cities of Ormoc and Tacloban lost their water supplies as the Leyte Metropolitan Water District in Pashana was clogged by fallen timber, boulders and sand. In addition the electrical transmission towers at the power station were damaged and 15 bridges fell. In Leyte the towns of Jaro, Julita, Dagami, Mayorga, Tabon-tabon and Dulag were damaged also. Three towns and two cities in Negros Occidental suffered damage.

It would seem, judging from local effects and the subsequent controversy, that the impact of cyclones, at least in the Philippines, is getting worse. It is likely that similar landslides and flooding, perhaps on a smaller scale, will result from future cyclones unless the real cause in terms of existing land-use practices, can be addressed. As the long term effects of deforestation and unsuitable land-use practices continues, with limited application of soil conservation techniques on destabilised slopes, such disasters are likely to become more frequent.

The Agricultural Base

The total land area of Leyte is approximately 800,000 ha; of this, 454,200 ha is disposable and certified alienable land. There is approximately 75,000 ha of irrigable land on the island. By 1985 about 9160 ha of this area was actually under irrigation, which may consist of national irrigation, communal irrigation or private pumps. The island has 76,725 ha of unclassified public forest and 269,332 ha of classified forest land (National

Economic and Development Agency, 1985; Regional Development Council, 1985).

Agriculture is very important to the economy of the Eastern Visayas region, contributing about 54 % of its gross product. Lack of diversity in export products exposes the island's economy to a high risk given its dependence on unstable world-market prices. In the Eastern Visayas region about 42 % of the cultivated land is under coconut plantations, with rice and abaca accounting for around 24 % of cultivated land area. There was also an increased tendency to plant sugar cane instead of food crops in the '60s and '70s (Jimenez and Franciso 1984, in Belsky, 1984). The world market for sugar has been unstable since the mid 1980s, and with a troubled coconut products market, has led to increased pressure on the land and a loss of subsistence crops, which has caused a deterioration in the state of land-based resources.

In the mid 1980s an attempt was made to reduce sugar production through a reduction in quotas from 78,000 Megatonnes to 52,500 Megatonnes between 1984 and 1987, with only 12,000 ha of land in the region to be cultivated for sugar.

The Philippines Government has made various attempts at agrarian reform, introducing concepts such as the Balanced Agro-Industrial Development Strategy (BAIDS) and in the Eastern Visayas Region, the *Kilusang Sariling Sikap* (KSS) and the *Kilusang Kabuhayan Kaunlaran* (KKK) programs, which were to focus on particular problems in agricultural production, rural poverty and loss of economic status (Regional Development Council, 1985).

One of the goals of BAIDS was to ... 'attain a stable supply of food and raw materials and maximise agriculture's contribution to the National Economic Recovery Program' (p. 4.2). While the names of such programmes changed with changes in governments, stated goals of agrarian reform remained much the same, as did the continuing failure to implement reform programmes (Pinches and Brown, forthcoming).

The KSS and KKK programs focussed on ameliorating problems such as low productivity of staple foods and the resultant shortfall in the supply of foodstuffs; inefficient marketing and distribution; lack of acceptance of improved technology for agricultural production; lack of co-ordination in research; and, perhaps significantly, dependence on a few traditional crops (p. 4.1). None of these programs appears to have had any far-reaching effect.

The Eastern Visayas, while devoting a large proportion of its land-based agricultural resources to production of export crops, was still contributing less than 2.5 % to the Philippines' gross domestic product, yet in 1985 consumption of domestically produced staples was expected to keep rising annually, while more land was probably required to produce greater volumes of export crops to offset falling world prices. Food staples in Leyte consist of rice and corn, meat, eggs, fish, and other commodities such as rootcrops, vegetables, fruits and cooking oil. The island may just be meeting its internal demand for meat, but it is unlikely that the production shortfall for rice and corn can be met – in Philippines terms this means that the nutritional requirements of the population are not satisfied and that some will become (more) malnourished (National Economic and Development Agency, 1985).

The main export crops grown in the region, apart from sugar, are rice, coconut, abaca, coffee and cacao, though amongst the upland farmers there is an overwhelming dependence on coconuts. A brief profile of Leyte's upland farmers is provided in Lumley and Stent (1989), as follows:

> Some of the upland farmers till lowland blocks of land. However, most cultivation for cash crops is of coconut plantations and abaca, with some cacao, coffee and peanuts being grown. Corn and rice are grown for cash and subsistence, while the main subsistence crops are sweet potatoes, cassava, mung beans, bananas and some other fruits and leafy vegetables. Farmers close to the coast also catch fish. Livestock kept are pigs, chickens and carabao (water buffalo). The average farm size per family is about one hectare. Farms are operated by share tenants, leaseholders, amortising owners and freehold owners. Share tenants may retain 33 % to 75 % of their produce. Products are often sold at local markets, but may also be sold direct to the consumer, to middleman merchants, to fellow farmers, or bartered for labour (p. 3).

Bearing in mind the financial dependence of Leyte farmers on world markets for a range of products lacking sufficient diversity, it is interesting to look at what has happened to the markets, and to levels of production of commodities produced for export on Leyte in the critical years between 1980 and 1985. The only products relevant for Leyte for which the Philippines is a significant producer are coconut and sugar.

Table 3.1 shows that in the six years from 1980 to the end of 1985 the overall value of the world coconut oil market fell by 1.3 % while the overall volume traded rose by 2.7 %, yet the value of exports from the Philippines fell by nearly 40 % and the volume exported fell by 29 %. The Philippines is the largest exporter of coconut oil in the world, yet the value of its market share fell from 67 % to 41 % of the world market value and its volume of market share fell from 82 % to 56 %.

Table 3.1 Philippines share of the world coconut and sugar markets, 1980-85

Year	Phils value US$000	World value US$000	Phils product (t)	World product (t)	Phils share (%$)	Phils share (%t)
Coconut oil						
80	566,848	842,526	918,521	1125,119	67	82
81	533,466	826,571	1039,900	1403,799	64	74
82	401,026	669,236	921,237	1296,515	60	71
83	515,812	714,535	998,252	1295,210	72	77
84	580,241	1077,087	587,575	1048,708	54	56
85	347,377	831,393	680,605	1155,788	41	56
%diff to 85	-38.7	-1.3	-29.2	+2.7	-26	-26
Sugar						
80	624,034	15,779624	1,746822	27,222032	4.0	6.4
81	566,560	16,005738	1,245441	28,689619	3.5	4.3
82	416,028	12,188695	1,260678	29,546203	3.4	4.3
83	299,350	10,906890	973,310	28,135100	2.7	3.5
84	290,350	10,847340	1,101580	28,185320	2.7	3.9
85	168,660	9,280990	582,530	27,142720	1.8	2.1
%diff to 85	-73.0	-41.2	-66.6	-0.3	-2.2	-4.3

Source: Center for Research and Communication, 1988.

The picture for sugar was no better except that the Philippines share of the world sugar market was not large at this stage. Nevertheless the proportion of its share of the market value fell from 4 % to 1.8 % and its share of market volume fell from 6.4 % to 2.1 % (from Center for Research and Communication, 1988).

For the other export crops produced on Leyte the Philippines holds an insignificant market share, so that in effect the upland farmers are almost entirely dependent on the world coconut market for their financial well-being. For interest though we can look at what has happened to the world markets for two of the crops that add some diversity to the export agriculture sector. Between 1980 and 1985 world production in cacao rose by nearly 27 % while the real price fell by around 23 %. For coffee, world production rose by 10 % while the real price fell by 27 %. Thus the cash economy of Leyte appears to have become reliant on world markets whose prices have fallen for every significant agricultural commodity traded.

Contemporary Socio-Economics

The total population of the Philippines is about 68 million (Japan Environmental Council, 2000) and the island of Leyte accounts for about 57 % of the population of the eastern Visayas region, its inhabitants numbering 1,600,000. According to Worcester (1921) Leyte's entire population in 1903 was 357,641 people. In the years 1975 to 1980 the average annual population growth rate was 1.5 % (Regional Development Council, 1985). In the Eastern Visayas region approximately 78 % of the people are rural dwellers with the remainder living in towns and cities. Approximately 49 % of the population is female while 51 % is male. The urban areas have a higher male population while the rural areas have a higher female population (National Economic and Development Agency, 1985; Regional Development Council, 1985). The province of Leyte is the most densely populated in the region with more than 207 people per km^2, compared with Eastern Samar, the most sparsely populated province, which has a population density of approximately 72 people per km^2.

The Eastern Visayas region is poor compared with the nation as a whole, with an average family income of around 71 % of the national average. In 1994 per capita GNP for the Philippines was only $950 US (Japan Environmental Council, 2000) which gives some indication of the poverty levels experienced by Eastern Visayans. This disparity in income between the Eastern Visayas and other parts of the Philippines is probably the main reason for the pattern of out-migration from Leyte to other provinces and to Manila (Wernstedt and Spencer, 1967). Indeed, one researcher commented that out-migration from a particular barangay on Leyte was an observed trend among the youth of the majority low income families who could not afford to finish their schooling (Tubiano, 1988). This is probably characteristic throughout the densely populated parts of the island where work is not readily avaliable. Much of this out-migration is permanent, with those who leave returning only for significant fiestas and family occasions.

Average life expectancy in the region is about 59 years, well below the national average, which stood at 68 years in 1995 (Japan Environmental

Council, 2000). Gastro-intestinal diseases and respiratory diseases are the main causes of death. The crude death rate is approximately 10 per 1000 head of population per year, while the crude birth rate is over 33 per thousand. Rates of infant mortality and deaths in early childhood are high in the Philippines as a whole with the mortality rate of children under five being 53 per 100,000 births (Japan Environmental Council, 2000). More than 67 % of children in the Eastern Visayas are malnourished (Regional Development Council, 1985). Health facilities in the barangays are very basic, although most barangays have access to a rural health unit. These units usually have a resident midwife and some trained assistants, and some run vaccination programs. Access to dentists and doctors is sporadic and sometimes non-existent (Tubiano, 1988; Tutor, 1988; Reoma, 1988 and Binoya, 1988).

The Philippines has a good eductaion record, with the primary school enrolment rate for the country being 100 % and the secondary school enrolment rate being around 64 % (Japan Environmental Council, 2000). Nearly all the barangays on Leyte have access to some education at the elementary level and the literacy rates are surprisingly high, with a rate of 81 % in Southern Leyte (National Economic and Development Agency, 1985).

The availablity of water is quite good, most of it coming from local rivers and streams. However, most facilities are community based and very few households have piped water. The use of domestic electricity is becoming more common via supplies from the Leyte Electric Co-operative. Some barangays have street lighting and a few households have direct electricity supply.

The management of the economy on a social and household level is conducted by various means and is heavily dependent on credit arrangements. Usurers often make cash loans at high interest rates and may demand repayment in cash or in kind. At the other extreme, farmers often lend each other commodities and repayments are in labour. Informal credit and loan arrangements vary from area to area and repayment of goods in kind is usually dictated by the seasons and harvests, with interest and capital repayments being made at harvest time (Lumley, 1988; Lumley, 1997).

The informal economy has a far-reaching effect on the socio-economic welfare of individual families and on the whole community. Because of the financial poverty of many barangay dwellers, co-operative arrangements between farmers are invaluable. However, some usury practices may be exacerbating poverty to such an extent that the cycle of poverty and debt becomes irreversible. In some areas farmers are trapped in a situation where they must sell their crop to a single buyer, to whom they are often in debt. For example:

> During harvest the farmer sells his entire peanut crop to the suki buyer for cash (which is often required to repay debts) without retaining seeds to sow in the following season. The suki agent who buys the crop sells the farmer seed peanuts for planting. These are supplied for in-kind interest at the rate of half a sack for every sack provided. In turn, the farmer must sell his crop to the suki agent who provided his seeds. The suki agent also extends cash loans and credit to his clients when they have a crop sown, since he is assured of repayment at harvest (p. 5, *ibid*).

In summary then, Leyte, even by Philippines standards, is underprivileged. The incidence of poverty and malnutrition is high and the land lacks the capability to produce the goods required. It is also often poorly nurtured. Barangay life is by no means homogenous in terms of wealth or class, and there are examples of extreme wealth and extreme poverty at the one place. Access to education is reasonable, while access to health care and fair credit is poor. The social life of the barangay revolves around religious festivals and family events, and one of the most prevalent forms of gambling is on cock fights. Making a living is difficult and the risks of daily life are exacerbated by the often devastating effects of natural disasters, particularly cyclones.

Chapter 4

The Socio-Economic Survey

Introduction

The details of study site selection and the development of the socio-economic surveys resulting from the original project involving ACIAR, ViSCA, La Trobe University and the Victorian Department of Conservation and Environment, are presented at the beginning of this chapter. This project was the basis for the research conducted by Parrilla (1992) and for some of the research contained in this book. After a description of the establishment of the original project, a brief summary of Parrilla's (*ibid*) findings is presented. This is followed by an outline of the research proposals and surveys used in this book.

The Original Project

As discussed earlier, the overarching goal of the original project was to assess the reasons for non-adoption of known soil conservation technologies by upland farmers in Leyte. This was done in the context of a wide range of socio-economic surveys of selected Leyte upland farming communities, which had generated data that might help to explain farmers' decision-making processes.

There is a range of soil conservation practices used in the Philippines to mitigate soil erosion. Some of these are relatively cheap and require little or no financial assistance. These include mulching, crop rotation, contour farming, strip cropping, use of cover crops, use of perennial crops and minimum tillage practices. Some practices, however, are costly and require technical knowledge and so government assistance through grants, subsidies and education programs is needed. These practices include terracing, irrigation dams and forestry.

Parilla (1992) provides a comprehensive description of such practices, including sloping agricultural land technology (SALT), and of a range of soil conservation models.

The surveys used in this analysis were designed to identify the socio-economic constraints on adoption of improved cropping methods. As Parrilla (*ibid*) states:

> If a farmer is to adopt an improved farming system, he must first be convinced that it is profitable for him to do so. This requires that benefits accruing to him from the innovation exceed their cost. This applies whether the innovation simply involves the use of new inputs, such as improved seeds or fertiliser, or new cultivation methods, such as the use of contour banks, etc. There is however, a major difference between these two sorts of innovation. Innovations of the former sort can normally be expected to show a return within a single cropping cycle, while the latter type may take a much longer period before it pays off. In either case it may be that whereas the total benefits exceed the costs the farmer is unable to capture all the benefits, so that while the innovation is economic from the social or community point of view, it is unprofitable for the farmer. In other words, there are positive externalities to the innovation (p. 18).

At the outset of the project demographic profiles of each of the study sites were made. The study sites, selected on the basis of upland conditions and representative demographic and physical profiles, and originally assessed with several other sites through rapid rural appraisal, were: Pomponan in Baybay, Tabing in Tabango and Canquiason in Villaba (all in the province of Leyte), and San Vicente in Bontoc (in the province of Southern Leyte) (see Figure 4.1).

During the course of the study, complete enumeration and proportionate stratification were used as sampling methods. Initially, farmers were classified at each selected site according to the area of land that they cultivated in the uplands and whether or not, according to their own judgement, they used soil conservation techniques. Forty farmers from each village were then selected by proportional random stratified sampling for the full range of detailed household surveys. These farmers would be intensively surveyed during the eighteen months of data collection. A total of 160 sample farm households were selected in this manner for the study.

During the stratified sampling procedure six strata were identified. These were as follows:

Small holdings (1.01 ha and less) with no erosion control
Small holdings with erosion control

Medium holdings (1.01-3.0 ha) with no erosion control
Medium holdings with erosion control
Large holdings (>3.0 ha) with no erosion control
Large holdings with erosion control.

The series of detailed surveys was designed around 13 main headings for which information was to be gathered either from secondary sources or by direct survey. Some data were collected on a pre-determined regular basis (daily, weekly, monthly etc.) and some were once-off surveys. The main survey instrument headings were proposed as follows:

A.   Synthetic profile of the village (secondary data source)
B.   Brief description of the population (direct survey)
C.   Characteristics of the farmer and household (direct survey)
D.   General farm information (mainly direct survey)
E.   Farm enterprise descriptions (direct survey)
F.   Household income, on and off-farm (direct survey)
G.   Household expenditure (direct survey)
H.   Credit and loan arrangements (direct survey)
I.   Crop and livestock production and disposal (direct survey)
J.   On and off-farm labour profiles (direct survey)
K.   Family time allocation (direct survey)
L.   Erosion and disaster occurrence (direct survey)
M.   Market prices for commodities (direct survey).

Parrilla *et al* (1991) summarise the manner in which the surveys and the interview schedules were designed and collated, as follows:

> The project's interview schedule was designed in an iterative manner. The first-round survey was employed to gather basic demographic and socio-economic data to classify farmers for sampling purposes. In the second and subsequent rounds, the information generated was used to develop further detailed questions. Questions were grouped according to the kind of survey conducted and types of respondents ...
> The project used different methods of data collection depending on the type of information required, namely formal and informal methods. The formal methods were: single interview survey, frequent interview survey, monitoring, record keeping and direct measurement, while the informal methods included reconnaissance survey, informal follow-up interviews and participant observations (p. 4).

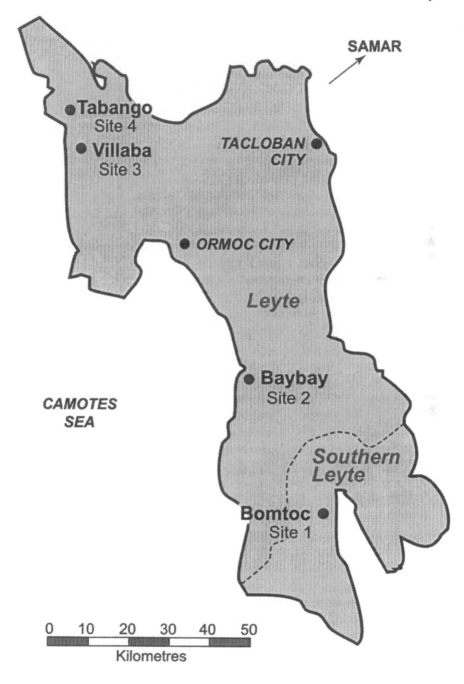

Figure 4.1 Location of the study sites in Leyte

All on-site interviews were conducted by research assistants from ViSCA. They spent weeks at a time in the field, living in the village to gain the confidence of the interviewees. A research aide was recruited at each site to help the research assistant dedicated to that site with data collection. Field work began in September 1986 and continued until June 1988. The detailed survey instrument used by research staff to gather the required data was divided into three parts which were implemented over the research period. Parts I, II and III incorporated the headings in A to M above.

Part I broadly dealt with demographic and general farm information. Part II was used to monitor cash flow, crop production and disposal, activity and soil erosion calendars, labour use and cash availability, while Part III was used to gather information detailing farmers' upland blocks. Implementation of Parts II and III commenced in March 1987 and ended in June 1988. By May 1988 there had been 10 field visits, and the research assistants had spent their time between visits at ViSCA encoding information collected onto computer databases.

In October 1988 a seminar was held at ViSCA. Here, project descriptions and preliminary results were presented for discussion. At this stage, very little of the massive database had been analysed.

Barangay Profiles

In order to gain a thorough understanding of the farmer decision-making process, and of the nature of the information that was gathered for the socio-economic survey, it was necessary to develop a picture of the study sites themselves. For this reason a short profile of each of the study sites is presented below. This, along with a general description of the influences on daily life in the area, will help to give some understanding of the conditions under which the people being assessed were living.[1]

General discussion   Each province in the Philippines is divided into a number of municipalities, and each municipality comprises a number of barangays (villages). Each study site surveyed in this analysis is a barangay. As discussed earlier, contemporary barangays have been derived from the early forms of settlement introduced by Indonesian and Malay settlers over a period of more than 1000 years. In those days the local leader was usually a Datu. In contemporary times barangays are led by barangay Captains who have been appointed or elected. Many barangays have *sitios* attached to them. These are outlying areas or hamlets of the barangay. Philippines' barangay culture derives from an extraordinary mixture of the cultures of preceding colonisers and immigrants to the archipelago. The immigrant cultures have been overlain to produce new cultures, which differ from area to area depending on the dominant religion, ethnic mix and other influences. Most barangays are predominantly Roman Catholic, derived mainly from the Spanish Catholicism brought to the Philippines by the Jesuit and Augustinian priests and monks who settled the islands while Spanish colonial administration remained in Manila. However, some significant cultural features in Leyte (such as St Valentine's day celebrations, the Seventh Day Adventist Church, and basketball) were introduced by the Americans.

Religious and social activities are closely linked, and they influence other aspects of barangay life. For some religious festivals farm labour is proscribed. For example, no labour may be conducted during Holy (Easter) week, from Holy Monday until Easter Sunday. In addition, all significant saints' days and religious festivals are celebrated with feasts which may involve killing valuable farm animals, such as pigs, even in times of hardship, and incurring debts to cover the expense. In most barangays there are weekly activities such as cockfights, where money is gambled on the outcome.

Apart from religious observance, other aspects of barangay structure and social behaviour seem to be derived, at least in part, from neither Spanish nor American customs. The various systems of co-operative labour and much of the in-kind economy may have come from earlier Indonesian and Malay traditions (this may also be true of the informal credit economy). Some forms of money lending and calculation of interest owed appear to be Chinese in origin.

Another aspect of the country's history which has a strong influence on contemporary socio-economics is the impact of land tenure and land ownership. In some areas there is an almost feudal system where much of the land on which barangays are located is part of a massive hacienda. The haciendas were tracts of land taken up by Spanish settlers, mestizos, and sometimes the hereditary overlords of the original barangays, when Spanish administration granted title to the land (as discussed in Chapter 2). Much of the land in and around barangays is often still owned by one or two families who have a great influence on local politics, labour and crop ownership, as well as on the informal economy. Nearly all cultivation of farmlands is by hard manual labour or use of animal power. The usual farm implements are the bolo (a large, curved knife used for a range of activities), the carabao-drawn plough and the thresher (Tubiano, Binoya, Reoma and Tutor, 1988).

When the study sites were first selected for the socio-economic survey, each was assigned a number for ease of recording. During the course of this discussion the study sites will be considered in the order of the numbers allocated to them, which are as follows:

Site 1 - San Vicente, Bontoc
Site 2 - Pomponan, Baybay
Site 3 - Canquaison, Villaba
Site 4 - Tabing, Tabango

Views of two of the study sites are shown in figures 4.2 and 4.3.

Site 1: San Vicente, Bontoc The barangay of San Vicente is located 3 km from the main town of Bontoc and 2 km from the closest market where produce may be sold. Transport to the barangay is good, with access by motorcabs, jeepneys and buses. The barangay population at the time the survey began was 701 people spread over 126 households, with an average of 5.5 members per household. The average household cultivated 1.7 ha of land and households were spread in isolated clusters around the barangay. Most of the residents were born in the area. The main water supply was from a natural spring and there was some irrigated land fed from pumps. Electricity was available for those who could afford it, although most households used kerosene lamps.

The barangay was reasonably well serviced, with a primary school providing education from grades 1 to 4, a health centre, a playground and a cock-pit, and Catholic and Seventh Day Adventist chapels.

The municipality of Bontoc occupies 8,912 ha while the barangay occupies 114 ha. The land is generally flat with slopes rising 50-100 metres above sea level. At the time of the survey the main cash crop was corn (70 ha), followed by rice (25 ha), pasture (12.6 ha), coconut (5.5 ha) and abaca (1.5 ha). Other crops grown in small areas for cash or subsistence included legumes (peanuts or mungo), coffee, cacao, root crops, bananas, tobacco, vegetables and fruit (star apple, santol, pineapple, jackfruit and citrus). Carabao (water buffalo), pigs, chickens, cattle, goats, dogs and cats were also raised in the barangay.

Land ownership status of the farmers was a mixture of tenants, leaseholders, mortagees and direct owners. Some tenant farmers paid a fixed rent to owners, regardless of the harvest yield; others paid a percentage of the harvest, ranging from 25% to 66%, to the owner. At the time of the survey six amortising owners in the barangay had paid off their mortgages and received title to their land.

The rice farmers had three crops per year and usually used IR (International Rice) 60 and IR 64 varieties. Usually pesticides and fertilisers were applied, and a range of labour practices, such as family labour, hired labour, co-operative exchange or contract labour was employed throughout the barangay. Labourers were hired with their work animals on a daily basis. Contract labourers were paid in-kind, at an average rate of one-fifth to one-sixth of the harvest (Binoya, 1987).

Site 2: Pomponan, Baybay Pomponan was the largest barangay surveyed, with a population of 3,105 people in 670 households and an average of 5 members per household. Transport was fair, with access by tricycle and jeepney. The barangay is 6.5 km from the main town of Baybay, and has 21 sitios. The main pattern of settlement is that of households clustered closely along the barangay roads. At the time of the survey the upland areas of the barangay were supplied by water from an open spring, while the lowland areas had a water pipe. There was electric lighting for those who could afford it, although most households used kerosene lamps. Educational facilities were good, with a primary school and a high school in the barangay. There was a health centre serviced by a midwife, and a Catholic chapel and a Seventh Day Adventist chapel. For recreation there were a basketball court and a stage which is used for various recreational purposes.

The average landholding per household was 1.8 ha and the total land area of the barangay was 925 ha with 465 ha of cultivated land and 165 ha of forest. The main crops were coconut (250 ha), rice (130 ha), abaca (40 ha), root crops, vegetables and fruit trees (10 ha each), corn (5 ha), cacao (5 ha) and legumes (5 ha). Carabao, cattle, horses, poultry, pigs, goats and dogs were raised in the barangay.

Many of the farmers in the barangay were the share tenants of absentee landlords. The %age share of each harvest taken by the landlord varied from 25% to 75%. On the lowland rice farms the varieties cultivated were IR 42, IR 60 and IR 64, while in the uplands Peta, Germas, Apostol and Dahili were sown. The lowland rice fields had a higher average yield than the upland fields, with 70 to 75 *cavans* (sacks) per ha compared with 55 to 60 cavans per ha. Most of the rice farmers used neither fertilisers nor pesticides, partly due to lack of cash, and partly from 'fear of poisoning', although both types of farmers used azolla to fix nitrogen in the soil.

The barangay was serviced by three private rice mills and a soybean mill. There were six copra buyers, three abaca buyers and various banana buyers in the village. There was serious water erosion in the hills, and bench terracing had done nothing to alleviate this (Reoma, 1987).[2]

Plate 4.1 A view of Site 1, San Vicente

Plate 4.2 A view of Site 3, Canquiason, showing *ipil-ipil* hedges in the background

Site 3 - Canquiason, Villaba   Canquiason had 300 households with an average of 4.5 people per household. The barangay is 11.5 km from the main town of Villaba. It has three sitios and is relatively isolated, the only usual means of transport in and out of the barangay being carabao. Houses are dispersed and facilities are limited, although there were both a primary school and a health centre at the time of the survey. The only religious building was a Catholic chapel and the only recreational facility was a basketball court. There was no electricity, lighting being by kerosene lamp. The only source of water was a spring.

The area of the barangay is 300 ha, with 200 ha being on hilly land. The average area of each household's landholding was 1 ha. The main crops were rice and corn, the subsidiary crops being coconuts, yams, bananas, cassava and camote (sweet potatoes). Livestock raised by the villagers included carabao, cattle, pigs, goats and chickens.

Most barangay residents derived the bulk of their income from farming. The majority of farmers were tenant farmers who gave 25% of the harvest to the landlord. Rice farmers used IR 54, 36 and 56 varieties and all rice crops were rainfed as there was no irrigation facility. Produce was marketed in the neighbouring barangay of Matag-ob. There were two rice crops and one corn crop per year (Tutor, 1987). The average farm size was 2.0 ha.

Site 4 - Tabing, Tabango   The population of Tabing was 1824, with 306 households and an average household size of six people when the survey began. The barangay has eight sitios. Although the roads in the area are rough, the barangay was well served with transport, there being jeepneys, buses and motor bikes providing access. The barangay had a primary school with grades from one to six, and a health centre with a staff midwife. The only religious facility was the Catholic chapel, while for recreation there was a basketball court. Drinking water was provided by an improved spring and electric lighting was available although most households used kerosene lamps. There was also a rice mill in the barangay. The average farm size was 1.1 ha.

Tabing occupies an area of about 800 hectares, 770 of which was part of two haciendas until the 1970s. Hacienda Salazar, which occupied 470 ha of Tabing, was included in an agrarian reform package in 1974. This led to the land being sold to the tenant farmers, who were amortising owners at the time of the survey. The 300 ha Hacienda Cabahug was still privately owned and was farmed by tenant farmers who usually retained about 75% of the harvest.

The main crops grown in Tabing were corn, rice, coconuts and rootcrops. Most of the upland rice comprised native varieties. Livestock raised in the barangay were pigs, chicken, carabao and cattle. The cattle were mainly the property of the two ranches in the barangay (Tubiano, 1987). Produce was sold in the markets of Tabango.

The history of Tabing is similar to that of many other barangays in the Philippines. Not long ago the entire area was owned by a Spanish mestizo family, the Velosos, who came from the island of Cebu (like many other settlers on Leyte). The Velosos acquired the land from Leyte's original settlers, who could not read and never gained title to the land they farmed. These farmers allegedly agreed to barter the lands they cleared and tilled for clothes and other goods (Tubiano, 1988). The Salazars, and the Cabahugs who own the remaining hacienda in Tabing, are descended from relations of the Velosos.

The Questionnaires

The survey questionnaires were designed and administered in such a way as to collect as complete a data set as possible on all the activities of the selected respondents, while gaining their confidence. Their confidence was required in order to elicit honest responses, especially about income, credit and land cultivated. This was done by ensuring that the research assistant who was administering the survey lived in the barangay while conducting interviews and collecting data, and by employing a person from the barangay to assist. In Pomponan, the site with an insurgency problem, there was initially some suspicion as the residents were fearful that the research assistant was a member of the NPA. However, he was able to gain the trust of the barangay residents. Because of the requisite completeness of the data set, a series of detailed interview schedules was developed. These were to cover the activities of every member of each co-operating household over the interview period. The field visits after respondent selection spanned the period from March 1987 to June 1988 for all sites.

The first questionnaire to be applied to each household constituted a comprehensive survey of every member with the aim of computing the socio-economic status of the household. This included the marital status, educational attainment, gender, years in farming, age, and primary and secondary occupations of the household head. For every other household member the following information was gathered: gender, age, primary occupation, secondary occupation, marital status and relationship to the household head. Information was also collected on organisational affiliations, participation in festivities, aspirations for self and family and reasons for farming at the site.

The next part of the questionnaire comprised a survey of general farming information including: area of upland and lowland holdings, animals raised, ownership of farm implements, practice and methods of soil conservation (including crops used for conservation) and reasons for not using soil conservation (O'Brien, 1989).

This comprehensive general survey elicited the information necessary to develop a profile of the respondents' positions in the strata classes used in the project, and to set the scene for the more specific surveys to follow. As the project progressed some additional surveys were developed to fill emerging information gaps. One of the additional surveys comprised a more detailed questionnaire than originally administered on credit received and extended. This was because it became clear from the earlier survey that it would not be possible to determine who was borrowing from or lending to whom. More information was also gathered on upland farming practices, weather and bad cropping years.

Early Project Findings

In October 1988 a workshop was held at the Visayas State College of Agriculture (ViSCA) in Baybay, Leyte. Here, papers on the background, research goals and preliminary results of the project were presented. As this seminar took place before completion of the project, very little analysis had been conducted and data collection was still being finalised. However in a paper by Lumley (1988), an attempt was made to assess patterns of income, borrowing and lending at the study sites. While this was only an interim analysis, and the results were inconclusive, some interesting points were raised and it was possible to assess the efficacy of some of the survey instruments. One survey instrument in particular (credit received and given) was found to be lacking in rigour, and was producing results inadequate for the requirements of the project. This is because the survey was very simple and did not generate enough information about the nature of borrowing and lending among the survey participants. Cash receipts were recorded, as were receipts of payments in-kind, quantities of product and product prices. This made it possible to estimate the value of cash and other borrowings for in-kind loans and repayment of cash and in-kind loans.

There was, however, no record of the duration of the loan, the schedule of repayments (lump sum or periodic), nor of the institutional status of the lender (bank, co-operative, usurer, or friends or family). In fact, some of the results from analysis of this survey were perplexing. We were able to identify which transactions were loans being made and which were loans being repaid, but we had no idea of who was borrowing from whom, nor of who was lending to whom. Lumley states:

> Other interesting information emerges from a quick appraisal of the data. The lending of cash appears to be spread almost randomly among respondents within a wide range of total incomes. As we would expect, only those with a relatively high income are lending large amounts of cash. Each site has at least one respondent from the highest income bracket lending in excess of one thousand pesos, except in Canquiason, where the larger cash loan (500 pesos) is made by someone with one of the smallest incomes (*ibid*, p.10).

One clear fact that arose from preliminary analysis of the data was that within each village there was a huge variation in income between the poorest respondents and the wealthiest.[3] In Canquiason, Villaba the respondent with the highest income earned more than 800 times the amount earned by the respondent with the lowest income. Because the respondents were selected by random stratified sampling it is possible that we captured the poorest and richest households in the village (i.e. those at the extremes of the normal income distribution). However, even if this were the case, the observations would still be valid.

When the results were considered overall, as a function of the total sample from the four sites, it was clear that income and farm classification, according to the six strata described earlier, were closely correlated. As expected the cultivators of the smallest areas of land generated the least income and the cultivators of the largest areas of land generated the most income. One of the aims of this analysis was to determine the statistical significance of farm size and soil conservation adoption, income, land tenure, and borrowing and lending arrangements as they relate to each other.

Sources of income as well as total income were very variable within and between sites. There were three possible income sources identified from the surveys: on-farm income; income from off-farm labour; and income external to farming (such as remittances from members of the family living away from home). It was found that Canquiason had the largest number of households deriving no income from the farm, while Tabing had the smallest number of households with no on-farm income. Access to formal credit also varied between the study sites:

> Of the four study sites, three (Tabing, Canquiason and Pomponan) have no direct access to formal lending institutions. In San Vicente the main

institution used is the Bontoc Credit Co-operative Inc. (BCCI). In Pomponan, farmers had access to the Rural Bank of Baybay through the Masagana 99 programme, but this no longer operates (*ibid*, p. 4).

The effective rate of borrowing from the formal source, the BCCI, was 12% per annum. In the three sites without formal credit arrangements the only recourse for those requiring a loan was to borrow from neighbours and family, money lenders or usurers. Unfortunately our original survey told us nothing about the motivation for the villagers to borrow money or products, who they borrowed it from, or under what conditions they were required to repay it. Lumley states:

> Informal credit and loan arrangements vary between the study sites. These range from cash loans from private money lenders at usurous rates, to repayment of commodity loans by labour. In many cases it is difficult to determine the annual interest rates since prices of goods and labour fluctuate according to climatic conditions, season and the market demand. Interest on loans from usurers is usually determined on a monthly basis and appears to be in the range of 10% to 20% per month. This is equivalent to an inferred rate of 120% to 240% per annum. Repayment of goods in kind is usually determined by seasons and harvests, with interest in goods being paid at harvest time (*ibid*, p. 5).

As a result of this preliminary analysis of the income, credit and loan data it was decided that a more comprehensive survey of borrowing and lending was required. Thus two new surveys were devised. These were survey instrument number 4.1, credit received, and survey instrument number 4.2, credit extended. These surveys were far more detailed than the original survey 3.5, and sought to identify the source of the loan, conditions of repayment, duration of the loan, amounts borrowed and repaid, and the stated purpose and actual use of the loan. Survey 4.2 covered credit extended and required similar information about the loans to that gathered in 4.1, but from the lenders' perspective rather than that of the borrowers.

After the project workshop in October 1988 there was no further formal presentation of analytical results (apart from a summary of the methodology and interim results in Lumley and Stent, 1989) until the presentation of a paper at the 35th Annual Conference of the Australian Agricultural Economics Society (Parrilla *et al*, 1991). This paper aimed to demonstrate the use of cluster analysis as a multi-variate statistical procedure in assessing the project results. The clusters were determined according to the location of discontinuity in dendograms constructed from the data. This related to adoption of soil conservation practices and the following criteria: family characteristics, farm characteristics, farmers' characteristics, and household characteristics and perception of soil conservation. Family characteristics included income, wealth, liquidity, labour contibution, family size and age of members. Farm characteristics included productivity, farm income, upland area proportion and upland farm size. Farmer characteristics included age, years of farming experience and education level. Household characteristics included all of the above and were related to the farmers' perceptions of what constitutes soil conservation.

This analysis did not include tests of statistical significance because it is not in the nature of cluster analysis to generate such results. However, the overall analysis produced some very interesting general conclusions. These were summarised as follows:

> 1. That net farm income, liquidity positions, land tenure, age and farming experience of the farmer are associated with the adoption of soil conservation practices in the upland areas. The greater the adoption of improved upland practices the greater the farm households' earning capacity. Also the older and more experienced the farmer, the greater is the adoption of soil conservation practices.
> 2. That land productivity is positively related to the adoption of soil conservation practices. Land tends to be more productive if soil degradation could be minimised. Farmers with highly productive farms and high adoption scores tend to have higher liquidity positions and net farm earnings.
> 3. That tenure status and size of the farms cultivated may have some influence on the amount of net farm income and net family income received by farm households per year.
> 4. That low net farm income households with low adoption scores tend to receive more off-farm income per year and higher propoertions of the family labour contribution to total on-farm labour.
> 5. That farm households characteristics such as net family income, liquidity positions, wealth positions, area of upland farm cultivated, land tenure, age and farming experience are associated with farmers' perceptions of soil conservation practices. It follows that positive farmers' perception and adoption of soil conservation is greater if they have high net family income, high liquidity positions and wealth positions. This indicates that those farm households with experienced farmers and have[ing] better earning capacity tended to perceive that practicing soil conservation is advantageous. Also, the larger the size of the upland farm cultivated and owned, the better their adoption and perception toward soil conservation.
> 6. That using the total sample mean of certain characteristics in identifying a typical farm, or farm household may not always be practical or reasonable (pp. 33-34).

In September 1991 O'Brien presented a paper to the IBSRAM workshop (O'Brien, 1991) in which results from the project were summarised. It was noted that the farmers appeared to demonstrate very high discount rates (p. 454), showing a greater preference for short-term returns over long-term returns. This is, in part, due to risk. But, as O'Brien states:

> In spite of its central role in much theoretical and applied economics, time preference has received little empirical attention. The most obvious reason for this is that with rational utility-maximising consumers and perfect capital markets, the time rate of preference is equal to the consumption rate of interest. These market interest rates are normally available to the researcher (p. 455).

In this research, original surveys 4.1 and 4.2 provide information about the interest rates paid by farmers and demanded by creditors, but the information is not sufficient to assess farmers' private time preference rates if, contrary to theoretical economics as applied to developed nations' economies, market interest rates do not accurately reflect time preference rates. This will be further discussed.

Parrilla's Research

In August 1992 a PhD thesis which drew on the database generated by the original project was submitted (Parrilla, 1992). In this thesis the work using cluster analysis to assess farmers' decision making (see Parrilla *et al*, 1991) was further developed. The thesis was very broadly based and gives a useful overview of village farm level activity at the study sites, although given the scope of the thesis there was little opportunity to analyse the differences in decision making and socio-economic conditions between sites. The general objectives of the study were stated as follows:

> (1) to provide a description of the upland farm households and their farming systems;
> (2) to determine analytically whether or not it is beneficial for farmers to adopt improved soil conservation practices;
> (3) to identify and examine those factors which stimulate or inhibit the adoption and use of soil conservation practices (p. 204).

These objectives were achieved in the course of the study. The description of the farm households and farming systems (1) is comprehensive and provides a good deal of insight into the socio-economic conditions and farming practices at the study sites. It was concluded as a response to (2) that it is indeed beneficial for farmers to adopt soil conservation practices, but the constraints on their daily lives may prevent achievement of a socially optimal outcome under prevailing conditions. The constraints on the adoption and use of soil conservation practices were assessed through limited dependent variable regression, a Two-Limit Tobit model and cluster analysis. The most significant influences on the adoption of soil conservation techniques (outlined in (3)) were found to be as follows:

(i) The higher the *liquidity* of the household the higher the rate of soil conservation adoption.

(ii) *Off-farm employment* has a negative relationship with the scale of soil conservation adoption.

(iii) The greater the *slope* of the land the higher the rate of soil conservation adoption.

(iv) The *age and experience* of the farmer influences the level of adoption, with experience having a greater influence. Less experienced farmers practice less soil conservation.

(v) The greater the *recognition* of soil erosion as a problem, the greater the rate of soil conservation adoption.

(vi) The *rate of interest* on borrowed funds (which in Parilla's work is used as a surrogate for the discount rate) influences rates of adoption. The higher the interest rate on borrowings the lower the level of soil conservation adoption.

(vii) *Tenant farmers* are less prone to adopt soil conservation than owner operators.

A more detailed analysis of the role of land tenure, income and farmers' time preference rates relative to interest rates on borrowings has been conducted in the present study, in order to try and examine how and why these factors affect adoption of soil conservation techniques. This analysis is conducted in the context of the history and culture of the country and of the economic theory to be presented in Chapter 5, in order to assess whether the assumptions of economic theory in natural resource allocation are efficient and accurate, and to inform the overall environmental and social policy debate.

Analytical Proposals and Survey Development in this Book

The reason for concentrating on rates of interest paid on debt, and on the individual discount rates of farmers, as factors influencing the adoption of soil conservation practices, is primarily because the interest rates observed in earlier work with the study samples (see Lumley 1988) were so high. As stated before, interest paid on debt is sometimes used as a shadow for an individual's discount rate. Yet it was hard to believe that a farmer's discount rate could be as high as 250 % even if poor, upland Leyte farmers' private time preference rates were much much higher than those reflected in more developed countries (often assumed to be around 10 %). However, if farmers' private discount rates were extremely high then it would not be surprising that they were failing to adopt soil conservation practices. This is because, in theory, the higher an individual's discount rate, the greater that person's preference for present consumption. Because the benefits of techniques like soil conservation might not be realised for several years, farmers with very high discount rates were deemed to be unlikely to adopt them.

In this research, the individual discount rates of farmers were investigated in the light of interest paid on debt, the purpose of their loans and the effect on discount rates of changing the circumstances for which money was borrowed. This was done to assess the effect of these factors on farmer decision making.

In Chapter 5 some assumptions in economic theory relating to discount rates and the use of discounting are discussed, since economic theory is often used to inform and influence social and enevironmental policy. The perceived discount rate survey was designed to assess some of these economics assumptions and these were tested in a series of hypotheses presented in Chapter 7.

In this work the standard-bidding approach for inferring private time preference rates (PTPRs) was used to establish baseline PTPRs for the upland farmers. Willingness-to-pay (WTP) questions were also used to assess PTPRs. The soundness of this approach was examined by comparing WTP PTPRs with standard-bidding PTPRs and testing the comparison statistically   The PTPRs stemming from the discount rate survey are referred to as the perceived discount rates. The results of the analysis have been used to propose policy based on the factors influencing farmer decision making, rather than on theoretical assumptions, as is often the case with policy development.

Research proposals for this book were outlined in Chapter 1. A survey additional to those conducted in the original project was required to determine the discount rates of the respondents as proposed for this book, since none of the original survey instruments could be used to estimate the private time preference rate of individual farmers. While assumptions in economic theory generally use market interest rates as a shadow for borrowers' discount rates under assumptions of perfect capital markets and freedom of choice, it became clear that these assumptions do not hold in the rural market systems operating in less developed nations such as the Philippines.

A more detailed discussion of discount rate theory and its role in resource allocation will be presented in Chapter 5. However, because earlier results indicated that discount rates probably have a significant role in influencing the non-adoption of soil conservation practices (especially among low income earners) and hence optimal resource allocation, some results from an empirical study conducted in India will be briefly discussed here. This will be done to introduce and rationalise the development of a further survey which was devised to elicit more specific information about time preference rates of the co-operating farmers in the Leyte project.

The research on the measurement of time preference rates (discount rates) in rural India (Pender and Walker, 1990) set out to estimate the discount rates of household heads in two rural villages in semi-arid tropical India. This was done using experimental games and hypothetical questions and the results indicated that the mean discount rates were 30 % to 60 %, which was *higher* then the average interest rates on outstanding debts. (In the case of Leyte, we would expect farmers' discount rates to be *lower* than rates on debt outstanding because interest rates on some in-kind transactions may be as high as 240 to 250 % (see Lumley, 1988 and O'Brien, 1991).) The Indian results showed that wealth was significantly inversely proportional to discount rate, and that discount rate was also influenced by the magnitude and duration of the loan. The effect of magnitude and duration of the loan was of particular interest as the few empirical studies that had then been conducted (e.g. Loewenstein, 1988) showed results that were sometimes inconsistent with predictions based on neo-classical theory (Pender and Walker, 1990).

It is clear from the results of the case study in Leyte, and from Parrilla's work, that several factors influence the adoption of soil conservation technologies, and that some of these bear a direct relationship to an individual's discount rate (such as wealth position, income, age, borrowing behaviour and education). Because an enormous amount of observed data had already been collected and analysed, and because of the well established network of farmer co-operators who provided this information over a number of years, that information was tested against responses to new survey questions to assess farmers' real discount rates in relation to the interest rates they paid, and the other factors mentioned above.

As discussed earlier, there was great heterogeneity in wealth, income, age, education and experience within and between the study sites in Leyte. Because of previous analysis of the data (see, for example, Lumley, 1988;

O'Brien, 1991 and Parrilla, 1992) the wealth and income positions of all the respondents was known, as was who was borrowing from whom, how much they were borrowing, what they were borrowing (in-kind), under what conditions commodities were lent and the duration of such loans. The nature of the relationship between time preference rates and adoption of soil conservation technologies have been analysed, and assessed for whether results were consistent with other empirical studies. For this reason the new, hypothetical, discount rate survey was developed and it is used in conjunction with existing data to form the empirical basis of this research.

Development of the Discount Rate Survey

The goal of the discount rate survey was to measure the time preference rates of farmers and to see whether they varied with different borrowing conditions and purposes. As far as possible questions which would elicit a highly emotional response (such as those relating to borrowing money to save the life of a child) were avoided because such responses, while having a valid place in utility theory, would be difficult to factor into the analysis, containing as they do contingencies wildly beyond the scope of the theoretical market being assessed.[4]

The survey was designed to cover a range of issues which might affect the time preference rate of individual respondents. These were initially grouped under a range of headings as follows:

(i.a)   Borrowing a large amount in cash
(i.b)   Borrowing a large amount in kind
(i.c)   Borrowing a small amount in cash
(i.d)   Borrowing a small amount in kind
(ii)   Borrowing to avert loss or forfeit
(iii)   Borrowing to make a gain or profit
(iv)   Borrowing comparisons
(v)   Loss and gain comparisons
(vi)   Deferring receipt of money
(vii)   Lending
(viii)   Imposition of very high interest rates.

The effect of the duration of the loan was to be tested in (i.a) to (i.d) where questions relating to small and large amounts of cash and kind were posed over a range of effective discount rates from around 5 % to 110 % for periods of one year, 5 months (to fit in with local cropping cycles) and one month. All the in-kind questions were posed for rice *or* corn (depending on the study site; corn was the staple at one site and rice was the staple at the other three). The specified purpose of the in-kind loan was for seed, to reflect normal in-kind-for-interest lending circumstances. One question in this series related to eliciting a time preference rate for borrowing an enormous amount of cash (beyond the realm of possibility) over one year, to test the response against the other questions.

Category (ii) and (iii) questions were designed to infer whether the farmers' time preference rates would vary according to whether the money was being borrowed to prevent loss or to attract a financial gain. In these categories proposed loans were for one year.

Questions under category (iv), borrowing comparisons, were designed to provide a check on the effect of loan duration on the time preference rate. Under (i) respondents had been provided with 14 options to select for time preference rate over a range of durations. In (iv) respondents were asked which of a range of 8 options they would prefer at an effective fixed interest rate of 100 % (though the rate was not made explicit) for the same time periods as in (i). The respondents could make the choice of cash or kind loans (of equivalent value).

Category (v) questions again dealt with assessing how the time preference rate might be affected by borrowing for loss or gain (see (ii) and (iii)). However, in this case the respondent was asked to choose between borrowing to make a gain on the farm while incurring a loss of equal value *or* borrowing to avert a loss which would mean forgoing a gain of equal value. There were four possible answers to choose from, with only one choice. Two options related to financial borrowings, and two options related to in-kind borrowing. Again the effective (but unstated) discount rate was about 100 % per annum and proposed loans were for one year.

Category (vi) questions were of the standard format for assessing individual time preference rates. That is, hypothetical questions relating to deferred receipt of money were posed. Respondents were asked to indicate the smallest amount they would accept for deferring receipt of cash for one year so that their level of indifference could be inferred. One question was for deferring receipt of a small amount of cash and one for deferring a large amount of cash.

Category (vii) questions related to inferring time preference rates for lending compared to borrowing. These questions were devised to assess whether there was a difference between willingness to pay to borrow money and willingness to be paid to lend money. Again the comparisons were made over the same time periods as before (one year; five months and one month), and for small and large amounts of cash. Interest rates were effectively constant.

The last category, (viii), was designed to assess the effect of imposing very high interest rates when the borrower is in dire financial circumstances and cannot borrow from another source (which is a realistic scenario at the study sites). These questions covered two durations (one year and five months) and were developed to compare the individual's own time preference rate (as inferred from other questions, when he or she can make a choice about how much she or he will pay) with what they will pay when there is apparently no choice (an imposed discount rate).

Once the survey had been drafted, all category headings were removed, and questions were mixed randomly through the questionnaire to avoid presenting a series of themes and giving the respondent a cue about the nature and purpose of the questions. There was a total of 41 questions and they were nearly all designed to reflect local conditions as closely as possible (using local currency and local produce measures of realistic magnitudes). A copy of the discount rate survey questionnaire is presented in Appendix One.

The survey was administered to the identical respondents by some of the original research staff. Questions were delivered orally and answers were received immediately. The was no prompting by research staff, respondents were unable to compare their responses with other respondents at the time the survey was given, and there was no opportunity for respondents to calculate the effective interest rates and make comparisons with their earlier responses.

Notes

1. It is very important when conducting a socio-economic analysis, which aims to help explain the way in which people act, to have a broad understanding of all the factors that impinge on their decision making processes. Economic theory should not be applied in isolation from knowledge of the network of physical and financial constraints, cultural conditions, peer pressures and political realities that influence every person in the subjective and objective choices that they make.

2. It was discovered after site selection that Pomponan was situated very close to the local New Peoples Army (NPA) stronghold in the the nearby hills. The NPA presence had some effect on village activities. The NPA descended every so often for recruiting drives, public discussion sessions and sniping at the militia. During the course of the project there were several incidents which affected the life and livelihood of the barangay, including at least one shooting fatality and a manoeuvre which barricaded the residents (including the research assistants) in the barangay. There was some fear that the NPA would permanently occupy the village. The NPA presence in the hills continued for the duration of the project, and had two long-term consequences. One was that farmers with remote landholdings (particularly under abaca) stopped cultivating their blocks for fear of being ambushed or kidnapped. The second, possibly with wider implications, was

that all male villagers between the ages of 15 and 70 years were compelled by the barangay hierarchy to register for membership of *Puwersa Masa*, an anti-communist group. All such men were levied 15 pesos each and most of this money was used to form a vigilante group, the 'Viking Force'. The barangay was put under curfew and rifles and ammunition were purchased for the vigilantes. The villagers were forced to curtail their social activities and were frequently called upon to provide financial and other support to the Viking Force.

Another aspect of Philippines barangay life, which was reflected during the course of the surveys, was political corruption. In May 1987 the Congressional poll was underway. The Pomponan barangay and its captain were actively supporting a 'pro-adminstration' local candidate, yet the election was won by a Marcos loyalist who openly bought votes at 5 pesos each. A further incident reflecting the day to day hazards of barangay life occurred when one of the project's farmer-respondents was gored to death while trying to separate two male carabao who were fighting. The owner of the carabao which killed the man had sharpened its horns to a point to prevent it being stolen. In this region also, a serious and prolonged drought started in March 1987, causing a shortage of water which led to the irrigation pipes being guarded against water theft. In January 1988 a tropical downpour broke the drought. Fortunately the downpour did not cause much damage in Pomponan. However, the drought had caused hardship across the island and resulted in further erosion when broken by heavy rains.

While these incidents may sound extreme and dramatic, they illustrate the day-to-day adversity faced by many villagers in a developing country. Insurgency and counter insurgency, political corruption, drought, eroding downpurs and accidents are not uncommon, and have wide socio-economic effects that should not be ignored (*ibid*; Reoma, 1988).

3. This contradicts a popular idea that the socio-economic status of villagers is relatively homogenous.

4. The value of human life in general, and the value of children's lives to their parents in particular, is very difficult to accomodate within a standard economic framework. Nevertheless it is an issue that economists and philosophers should further consider, especially given the terrible scale of some natural disasters. However, considering the social, financial, cultural and emotional value of human life is beyond the scope of this analysis.

Chapter 5

Interest and Discounting

Introduction

As discussed in the previous chapter, a survey was developed to assess the perceived discount rates of farmers in a situation where they were offered some choice in accepting or rejecting the interest rate on borrowings offered to them. This was done, in part, to examine the validity of policy makers' applications of interest rate and discount rate theory to policy development. Earlier, two surveys had been administered to determine the actual borrowing and lending activities at the study sites. These surveys had covered credit received and credit extended. Information on the purpose of the loan, the amount, duration, cash or kind, source or destination and actual use of the loan was collected.[1]

Observed rates of interest on loans from usurers were very high and many farmers had no access to formal credit (see Parrilla, 1992). Theoretically, the farmers were operating in a free market, but it would seem that to many the consequences of not accepting the very high interest rates offered to them would be, by Western standards, extreme. For example, when the purpose of loans was analysed it was found that many farmers were borrowing to buy food.

The discount rate survey (conducted with the same respondents but in 1993) should reflect more accurately than the earlier surveys the farmers' actual private time preference rates, and should offer some insight into their reasons for not adopting cultivation practices which would appear to afford them long term benefits. In order to explain the effect of interest and discount rates on decision making, and to examine the nature of credit markets in the Philippines, discussion of interest and discount rate theory is warranted.

Discounting and the Private Time Preference Rate

Using standard economic economic analysis (such as cost-benefit analysis), governments choose between public projects by comparing the benefits and costs of each option. The streams of benefits and costs are then subjected to a discounted cash flow analysis over the projected life of the project, so that the net present value (NPV) of each option can be compared. Usually, when all possible benefits and costs have been accounted for, the project with the highest NPV is chosen. The practice of discounting means that the present value of current benefits is larger than the equivalent amount, in constant money terms, that will be received in the future. The further into the future the benefits accrue, the smaller their value in present terms. The discount rate chosen strongly influences the value placed on the future in the present. There is a large body of theory relating to choice of an appropriate discount rate, and measures such as the marginal opportuniy cost of capital (MOC), the long term bond rate and the social time preference rate (STPR) are often used (e.g. see Mishan (1975); Lumley, (1988 (b)); Lumley, 1997; Dasgupta and Pearce (1972)). Discounting is also used for making private investment decisions.

In simple neo-classical terms, for individuals making investment decisions (Krutilla and Fisher, 1975, p. 61):

> the discount rate was determined in the market by the interplay between individual time preferences and the productivity of investment. Individuals making their decisions regarding present consumption and savings would be governed by their time preference regarding consumption.

Thus the private time preference rate is used as the discount rate for assessing private investment decisions. The private time preference rate (PTPR) reflects a persons preference for present consumption over future consumption. PTPR is a subjective measure expressed as a percentage per annum and is usually inferred by questioning individuals about the amount of money they would require in order to defer receipt of a sum of money now, for one year. This technique was used to elicit the PTPR responses analysed in this research; it is referred to here as the standard-bidding technique.

The higher an individual's time preference rate the greater their preference for present consumption. PTPR is influenced by a range of factors including income, wealth, education, age and health. The poor, the old, the sick and the under-educated usually have a higher PTPR than the rich, the young, the well and the educated. Farmers in the Philippines could be expected to have a higher PTPR than, for example, young Western professionals.[2]

Interest Rates, Discounting and the Study Sites

In developed capital markets the marginal time preference rates of individuals participating in the market will approximate the market rate of interest, which is why the interest rate paid on outstanding debt might be used as a shadow for the discount rate used by private decision makers. This is because, in theory, the market interest rate on borrowing and lending is common to all borrowers and all lenders and moves freely in a competetive market (George and Shorey, 1980).

Thus, again according to theory, if an individual's time preference rate is below the market interest rate they will lend money and if it is above the market interest rate they will borrow money (*ibid*, p. 46):

> Consumers and producers will adjust their consumption and investment activities and their borrowing/lending requirements until their rates of time preference and the rate of return on investment are brought into equality with the market interest rate. Any change in preferences or investment opportunities will precipitate interest rate changes to maintain this equality ...
>
> In a perfect capital market, therefore, where each individual is confronted by the same rate of interest, each person will adjust his borrowing and lending, so that in equilibrium the marginal rate of time preference of all individuals will be equal.

This theoretical assumption holds reasonably well, in most circumstances, in the capital markets of developed nations. However, although it is widely accepted in policy development, it is not certain that the assumption holds in the village economies of less developed nations.[3]

If the assumption concerning capital markets does not hold, it may be because the markets are not operating efficiently. One reason for this may be that the conditions requiring competition are not fulfilled. This could be caused by the existence of monopoly in the credit markets.

Monopoly is of particular interest because some informal credit arrangements at the Philippines' study sites do appear to involve monopoly (for cash or kind), which may help to explain the apparently high market rates of interest and the very high perceived discount rates of farmers. As Galbraith (1987, p. 111) states:

The monopolist extended production not to where an impersonally determined market price just covered marginal cost but to where in consequence of his reduced price in general, his more rapidly falling marginal return just covered the added cost. That was where profits were maximised. No one could say that this was the socially optimal production and price. Production was at a theoretically smaller output than the competitive equilibrium. The price was higher.

It is not clear how great an influence monopolist credit suppliers have on the overall credit market (credit in-kind is an important commodity) at the study sites.

In Canquiason, Villaba, a system known as doblehan is applied to borrowing seeds for planting. Interest in kind is 100 % (paying back double what is borrowed) over the harvest period, which is about 5 months. In Tabing, Tabango, a monopoly situation exists whereby a peanut farmer must sell his crop to a single buyer and then must borrow the seeds for his next crop from the same buyer at usurous rates. Farmers in this system are not permitted to retain seeds from their last crop to plant their next (Lumley, 1988 (a)).

Only 62 % of the 157 farmer co-operators in this project obtained credit and many of these borrowed such small amounts that it would make little difference to their long-term viability as farmers. In fact '77 % of upland farm households are considered poor, with very limited resources to invest' (Parrilla, 1992, p. 150).

It could be possible that the cost (and price) of credit is so high as to preclude all but the more needy farmers from accepting credit, *should it be offered*. There may then be two classes of non-credit users: those who can manage their farms and subsist without incurring debt, and those who are considered uncreditworthy. Parrilla's results give no clear indication of exactly how use of credit is determined. Further analysis of less aggregated data is required to investigate differences within and between sites, for it seems, from the significance tests in Parrilla's work that it is those with *better* wealth positions who do not use credit. This again suggests that credit users may be in a position of little choice. It may be that the less creditworthy are charged the highest interest rates, if they can borrow at all. In Chapter 7 a series of hypotheses concerning rates of interest on cash and kind loans, and on willingness to pay for credit according to the purpose of the loan, are tested.

This research is concerned with examining why many upland farmers in the Philippines make the decision not to invest in soil conservation technology. Given the above discussion on interest and borrowing, the decision to invest would be based on the expectation of receiving a rate of return on investment which is higher than an individual's time preference rate. An individual will only borrow for consumption if the interest rate is lower than the individual's time preference rate. At the outset of this research the time preference rates of the farmers were not known, although neo-classical theory suggests that their PTPRs approximate the interest rate paid on debt. However, this assumes Western-style capital markets at the study sites, and given the extremely high interest rates it is doubtful that this assumption would hold for the investment decisions of the farmers. This leads to two parallel questions being posed:

- Does the interest rate on debt approximate PTPRs?
- Is PTPR an appropriate tool for determining investment decisions in natural resource allocation which affects all of society rather than just the individual making the decision?

Interest Rates and Private Time Preference Rates

The discount rate survey was developed to examine the actual time preference rates of farmers at the study sites in order to see whether they equalled the market rates of interest and to assess what factors influenced the private time preference rate. The implications for theory and policy, should the interest rate *not* equal the PTPR at the study sites, are now briefly assessed.

If the mean PTPR inferred from the discount rate survey were less than the mean rate of interest paid on loans, then we might conclude either that the respondents to the survey were not operating in an efficient capital market, or possibly that the loans were made in circumstances that were not reflected in the questions posed by the discount rate survey – or both. For example, a borrower might accept a high rate of interest on a small loan that was needed to avert *immediate* disaster, or for consumption that could not be deferred, yet in happier circumstances might require less compensation for deferring consumption and hence might express a lower PTPR. If this were the case, then when respondents' PTPRs were inferred for a range of circumstances, they would vary according to the conditions specified. Further, if indeed it appeared that farmers were borrowing at high interest rates only to meet pressing needs, then they would borrow at those rates to adopt soil conservation only if they perceived the (deferred) negative effects of not borrowing for that purpose to be substantially greater than the (present) negative effects of not borrowing to meet pressing needs.

Under these circumstances, the question that needs to be answered in order to devise appropriate policy is:

- what interest rate should be offered to farmers to provide an incentive for investing in soil conservation?

Presumably one possible response to this question is that if they were rational, farmers would borrow at a rate of interest which approximated the rate of return they could expect from investing in soil conservation. However, the expected rate of return may be difficult to determine because apart from local and individual differences in productivity gains, it may involve assessment of the risk of cyclones (among other things). Another response may be that if they were rational, farmers would borrow at a rate of interest which was lower than their PTPRs. In either case, if the prevailing interest rate precluded the decision to adopt soil conservation, then adoption would not take place unless credit were subsidised.

Another theoretical issue which should be considered here is that concerning the approximation of willingness-to-pay (WTP) measures with willingness-to-accept-compensation (WTA) measures, when assessing the private time preference rate. Both of these techniques are used in the perceived discount rate survey, and they are used in different ways. In most circumstances WTP is used as a shadow for the basic PTPR, and the basic PTPR is usually elicited by the individual bidding to the amount of compensation they are willing to accept to defer consumption of the money they are being offered now or in the future. The soundness of this assumption is tested in Chapter 7.

However, WTP to assess the effects of the threat of a severe loss and the promise of a large gain is also compared, and assessed against the base PTPR. There is a theory known as equivalent variation (e.g. Mishan, 1981), which suggests that the smallest amount required by a person to be compensated for loss of a good will exceed the amount they are prepared to pay for the good to obtain it. (This theory of equivalent variation plays a role in the WTP-WTA argument when attempts are made to value environmental goods). Despite the theory of equivalent variation many economists still use WTP and WTA interchangeably.

Knetsch has challenged the commonly held belief that people's willingness to pay to gain a good will be the same as the amount they are willing to accept to lose the good they already possess, and states (ibid, p. 5):

> The main issue is that there are two measures of economic value – one for gains and another for losses – and while we have long assumed that they would yield essentially equivalent assessments, the empirical evidence suggests otherwise. There is no dispute that economic values of both gains and losses are measured by what people are willing to sacrifice.

This area of dispute is further developed by Tisdell (1991), who cites a series of economists (including Knetsch) as having demonstrated a disparity between WTP and WTA. While he points out that economic theory predicts little divergence, Tisdell suggests that using WTP rather than WTA can lead to conflicting policy advice. This is a particularly relevant observation since many policy makers base their decisions on accepted conventions. Subsequent to the work cited above, further research has been conducted on willingness to pay techniques (for example, Blamey *et al*, 1995; Blamey *et al*, 1999) and such work has raised more questions about the validity of WTP as a reliable measure of economic value. Blamey *et al* (1995) raised the possibility that the results of contingent valuation surveys may not reflect standard economic values, but rather, they might reflect respondents' behaviour as good citizens, who use the CV survey as a referendum on the environmental resource under examination.

Private Time Preference Rate and Social Time Preference Rate

As mentioned earlier, governments often use discounted cash flow analysis when assessing public projects. One derivation of the discount rate they might use is the social time preference rate (STPR). The STPR is often taken to represent the weighted mean of all individuals in society's PTPR, the weighting being according to the amount they are investing. (Neoclassical theory suggests that in a perfectly efficient capital mark*et all* individuals' time preference rates will be identical, but as Mishan (1975, p. 200) states:

> In a more realistic economy in which men can borrow at different rates of interest, according to their credit worthiness, and in which they can increase their borrowing only by offering higher rates of interest, their marginal rates of time preference will differ.

Generally the STPR is assumed to be slightly lower than all individuals' PTPRs because (among other things) society as a whole has greater expectations of longevity than any individual. However, whatever discount rate is used, one of the vexed questions in economics today is whether project or investment returns involving the use of natural resources should be discounted at all. For example the use of STPR does not take into account the fact that resource degradation and depletion may not impose a cost on any one individual whose PTPR has helped determine the STPR, but such depletion imposes costs on society as a whole, now and in the future.

The Role of Discounting in Natural Resource Use

High interest rates charged for credit suggest a high time preference rate for borrowers, whether or not such a rate is a true reflection of their individual time preference rates. As a result, projects which will yield benefits some time into the non-immediate future are not conducted because of the emphasis placed on the present by high discount rates. In the case of the upland farmers in Leyte, such long-term projects usually involve use of natural resources (such as soil conservation projects). Pearce and Turner (1990, p. 211) state that:

> Economic analysis tends to assume that a given unit of benefit or cost matters more if it is experienced now than if it occurs in the future ... it will be immediately apparent that, if discounting has a logic of its own, it will still create problems when applied to environmental issues. To see this, consider a development that yields immediate and near-term benefits but which has fairly catastrophic environmental consequences for future generations ... so long as the weight we attach to the future gets less and less the further into the future we go, which, as we shall see, is what discounting does, then the less important such catastrophic losses will be. In other words, discounting contains an in-built bias against future generations.
> ... discounting affects the rate at which we use up natural resources. The higher the discount rate – the rate at which the future is discounted – the faster the resources are likely to be depleted (p. 211).

Pearce and Turner's statements on discounting and natural resource use do not bode well for the Philippines and other less developed countries! Apart from the very high discount rates extant in the country, in the Philippines a large proportion of the population lives in rural areas (about 70 %), and 50 % of the workforce is employed in agriculture or a related field. These people rely directly on the land for income, and for subsistence crops, so the implications of a rapidly degrading natural environment are serious. In addition, the steep slopes, which characterise so much of the landscape, are rapidly being cleared for forestry and agriculture and this makes the country very susceptible to the effects of cyclones and other natural disasters (see Chapter 3). The social and cultural history of the country, with its overlaid systems of usufruct, colonisation, Spanish and Chinese mercantilism, alienation of the common lands, stringent private property rights and usury, and a system of land tenure almost akin to feudalism (see Chapter 2) exacerbate the policy problems in addressing the tandem problems of natural resource degradation and the cost of credit.

Because of the significance of discounting in determining natural resource use, it is worth analysing and discussing aspects of long-term natural resource projects in the context of economic evaluation. Forestry provides a good example in this context, and it is of great significance in the Philippines and other less developed countries (Barbier *et al*, 1994; Palo and Mery, 1996; Sponsel *et al*, 1996). Forestry has two interesting aspects relating to choice of projects, and the discount rate has a very strong effect in project selection. The forest resource is obtained from either old growth, natural forests, or from plantations, both of which have very long cycles to maturity. Old growth forests are at the end of the growth cycle in terms of harvestable timber and have significant financial benefits in the present and small financial costs in the future if a discount rate is applied to harvesting in the present. Conversely, plantation forestry on unforested land has significant costs in the present (establishment costs) and small financial benefits if future harvesting is discounted to the present. Kula (1988, p. 9) comments that 'In economic analysis of forestry projects there are three dominant factors: the choice of discount rate; the choice of project evaluation method; and the future price of timber.'

The same arguments could hold in economic analysis of soil conservation projects, except of course the price argument does not apply to one particular commodity but to a range: for example the price of the crop to be planted in the soil, the price of undegraded land and the price of assets saved from destruction by averting soil loss, floods and landslides, and the price of maintaining the livelihood of the farmers affected. Many of these prices will be intangible (will not exist on the open market), and the costs and benefits of such projects do not necessarily accrue to the same people (bear in mind that many of the farmer – co-operators are share tenants of absent landlords). Forestry projects are, in the main, publicly owned rather than privately owned enterprises; that is, the forests themselves are usually on crown or common land and are owned or managed by the government on behalf of the people of a country or state (for example, this is mostly true for the Philippines, Britain and Australia).

Kula uses the example of forestry projects to demonstrate the sensitivity of the net present value to discounting. If the net present value of a project is greater than zero it is deemed to be feasible, if it is less than zero (negative) it is often deemed to be infeasible. But this approach should only be applied if all values have been included in the analysis, and this is frequently not possible. However, what usually happens is that NPVs of the various projects are compared and the project with the highest NPV is selected. It is easy to imagine which would be chosen, either felling an old growth forest, or establishing a plantation on untreed land, if the only criterion was the net present value of the harvest.

Lumley (1983) conducted an analysis which compared the discounted equivalent annuities of four land-use activities for discount rates varying from 2.5 % to 12.5 %. The reason for the comparison was to assess the relative benefits and costs of land-uses which could be used to mitigate dry-land salinity, compared with the base case or status quo (cropping). The three alternative activities were permanent pasture, lucerne and agroforestry. The length of the agroforestry cycle was a minimum of 25 years to harvest. All the other activities could achieve some return within a year. What is of great interest in this discussion is the sensitivity of the annuity to the discount rate. The agroforestry equivalent annuity surpassed all others at discount rates of 2.5 %, 5.0 % and 7.5 % but at about 8 % onwards cropping realised the highest financial return. (Cropping was the activity which was contributing to salinity and soil degradation, but on financial grounds a suitable alternative to cropping would have to have equivalent or better returns.) No decision maker could assume that the interest rate for agroforestry investment would remain below 8 % for the duration of the cycle and so any venture into agroforestry under these circumstances would be very risky. Hence, it would be *financially* viable to allow land degradation and salinity to continue, using discounted cash flow analysis, if all project returns were to be discounted above 8 %.

This theme, of projects involving natural resource conservation being thwarted by standard project evaluation methods involving what is often a subjective and arbitrarily determined discount rate, is ever recurring in economic analysis.

It is ironic then, that using even standard benefit-cost techniques, the discounted net costs or benefits of the indirect or external effects of a project may lead to selection of projects with lower direct financial returns, *if such effects could be measured.* However, standard techniques used to assess net present value, internal rate of return and marginal opportunity cost fail to take into account indirect effects, other intangible costs and benefits, and externalities. In addition, intragenerational equity, as well as intergenerational equity, is rarely considered, as such techniques tend to take into account investment criteria rather than social criteria. We always return to the fact that discount rate is the single most influential factor in project appraisal, selection and decision making, whether by governments or by private companies.[4]

On the subject of the conflict between social and development goals, and conservation, Tisdell (1991, p. 61) states:

> Note also that when the government is using social cost-benefit analysis to choose between alternatives, use of the market rate of interest as a discounting device may be inappropriate. If the market rate is excessive from a social viewpoint, the government, by using it for cost-benefit analysis may ... be led to choose development whereas conservation is socially optimal. This is, of course, assuming that reconversion of developed resources to a natural state is impossible or extremely costly (p. 61).

In addition, with specific reference to less-developed countries, Tisdell (*ibid*, p. 77) comments:

> Discount rates (interest rates) in less developed countries are often very high or effectively so. When rates of interest are high ... it is often profitable to liquidate slow growing populations of marketable living resources, such as slow growing tree species like teak. When the costs of liquidating such assets is the same, one might expect them to be realised more frequently in countries with high discount rates.

These comments are in keeping with the earlier discussion about forestry. Kula (1988) proposes a modified discounting technique for public projects in order to overcome some of the drawbacks of standard discounted cash flow analyses. This method involves estimating the value of a project to those who receive the net benefits arising from it. This is also known as the sum of discounted consumption flows. Kula (*ibid*, p. 83) states:

> Persons A, B and C, being alive at the precise moment of decision-making, become aware of the fact that they have benefits stored for them in the future ... By discounting at their subjective time preference rate, they bring those future benefits back to the time when they were made aware of what was lined up for them in the future. The same procedure is repeated for each beneficiary, and an overall figure is obtained by summation. This figure indicates the worth to society of the consumption stream created by the public project. If this figure is positive, the project is worth undertaking.

This approach has two major drawbacks:

- It still relies on discounting
- Although Kula has taken care to present a model for determining the social value of the consumption stream, in reality the technique would be very difficult to conduct.

While many economists recognise the resource-allocation efficiency and equity problems brought about by standard economic approaches to project appraisal and decision making, there have been few practical proposals, or analyses of exactly what basis decisions are made on, and how this diverges from theoretically held views. Clearly, assumptions in theories of 'level playing fields', freely competitive markets and perfect knowledge are not in keeping with reality. Interest rates have a strong role in determining the decisions of individuals but the determinants of interest in different credit markets are not well understood and, especially in less developed countries, nor is the way in which credit is made available. In addition, nepotism, monopoly and high-level corruption often have far more influence than economic theory in determining the allocation of natural resources. Policy makers need a much better understanding of reality than can be provided by economic theory alone. This can be assisted by developing some knowledge of the social and economic history of an economy, as well as through investigation of contemporary practices.

More about Discounting Theory

Choosing the appropriate discount rate, and indeed the use of discounting itself, in economic analysis continues to be controversial for many of the reasons raised so far (Lumley, 1997; Lumley, 1998 (a); Moseley, 2001). The role of interest and the part it plays in neo-classical economics in inferring discount rates of society and of individuals has been discussed briefly. We have looked at the possible effects of discounting and interest on decision making and at the manner in which this subjective concept can affect the allocation of natural resources. Because of the significance of discounting on decision making, resource allocation and inter- and intra-generational equity, it is worth looking at some of the theoretical aspects in more detail. Part of the concern about the use of discounting reflects acknowledgement that conventional market systems are not able to capture or allow for all the risks, uncertainties and imperfections in knowledge that adhere to long-term decisions, especially those involving the allocation of natural resources. In order to recognise formally some of the these difficulties Australian State and Commonwealth governments have agreed to use a concept known as the Precautionary Principle to guide decision makers. In the 1980s the German government formally endorsed this principle, and it has also been included in the Maastrich Treaty (Young, 1993).

Young (*ibid*, pp. 11-12) describes Australia's definition of the Precautionary Principle in its inter-governmental agreement on the environment as follows:

> Where there are threats of serious or irreversible environmental damage, lack of full scientific certainty should not be used as a reason for postponing measures to prevent environmental degradation. In the application of the precautionary principle, public and private decisions should be guided by:
> (i) careful evaluation to avoid, whenever practicable, serious or irreversible damage to the environment; and
> (ii) an assessment of the risk-weighted consequences of various options.

Young (*ibid*, p. 23) argues that the precautionary principle can be used to overcome the problems caused by traditional economic evaluation techniques which ... 'can cope with risk but not the concepts of uncertainty, ignorance and indeterminacy', and '... that assume individual consumer values equate with citizen based values'. Indeed, in 2000, the Australian Government included the precautionary principle in a new Act of Parliament designed to protect the environment (Commonwealth of Australia, 2000). In this Act (*ibid*, p. 388), the precautionary principle is that ...

> lack of full scientific certainty should not be used as a reason for postponing a measure to prevent degradation of the environment where there are threats of serious or irreversible environmental damage.

The second problem that Young raises above really appears to relate more to the actual choice of discount rate and the theory upon which it is based. The concepts of private and social time preference rates and their relationship to each other have been briefly discussed here. Presumably Young's individual consumer values reflect the private time preference rate, while citizen based values reflect one of the social measures (for example the social opportunity cost of capital or the social time preference rate). These concepts link with another of Young's stated objections to conventional evaluation techniques (*ibid*, p. 23), that they ... 'use algebra which assumes that society consists of a single person rather than an assemblage of people of varying ages'.

This issue has already been dealt with briefly when considering Kula's modified discounting approach in the previous section of this chapter. Young (*ibid*, p. 23) makes two other criticisms of standard techniques which are widely recognised and which have already been raised briefly in this chapter, that they ... 'assume that relative prices will remain constant over time and ignore the likelihood that the relative price of natural capital will rise over time' and that they ... 'use a constant discount rate which unrealistically implies that we expect the world's economies to grow exponentially for the entire period of the analysis'. The problem of understanding the validity of economic theory and the consistency of its application, is compounded by the fact that not all economists, analysts and policy makers appear to know of the requirement to use a single, constant discount rate (Pearce, 1971; Hussain, 1998; Lumley, 1998 (b)).

In response to criticisms of intergenerational inequities inherent in the use of discounting, whether we are referring to natural resource projects or any other government project, some economists have argued that governments have an obligation to future generations which transcends the selfish interests of current generations. Current generations may limit their own concern for future generations to their heirs, and this tendency may only reflect their personal time preference rates, so that there is a clear divergence between private and social discount rates. Thus, the argument goes, with each individual's contribution to the welfare of future generations being limited to a small personal contribution, a different collective (social) rate, with mass concurrance, can better take care of the interests of future generations. This, then, *may* lead to a distinct discount rate for social projects.

According to Arrow (1977, p. 420):

> the optimal solution would be to lower the required rate of return on all investment, private and social, for example by lending to private business at a lower rate than the market, or by driving the rate of interest down through a budgetary surplus or debt retirement. More private investment would be undertaken, so that the marginal investment would have a lower rate of return and the opportunity cost of capital would be lower. Then, without changing the rule of discounting the benefits of social investment at the opportunity cost, the interest rate would be lowered.

This is a circuitous and questionable argument, especially as we are still binding the social discount rate to capital markets. Krutilla and Fisher (1978, p. 63) express a similar approach citing Eckstein's 'second best solution' (Eckstein, 1958, p. 101) that if public resource projects are evaluated with a lower rate of discount, then the same rate should be used for discounting 'opportunity returns' in the private sector so that:

> The resulting difference in the present-value sums obtained by using the market, as compared with the social rate of discount in the opportunity sector should be reflected in the investment criterion governing the resource conservation program. This procedure would avoid the inefficiencies inherent in the two-tier discount rate and perhaps meet the conservation objectives ... by favouring alternatives with a higher futurity of yield.

Arrow (*op cit*) concludes by stating that using a social time preference rate rather than a private rate is a matter of value judgment.

Several issues have been raised in the foregoing discussion but the main points of contention are whether government and /or natural resource projects should attract a different (lower) discount rate in project evaluation and whether this should be based on an STPR. Most economists seem to accept that an opportunity cost discount rate and a time preference discount rate diverge when the assumption of perfect capital markets fails. Many considerations (including personal and emotional ones) contribute to the determination of an individual's time preference rate and hence, given that society is a collection of individuals, to an STPR.[5]

Implications for Policy Development

Given the far-reaching effects of discounting practices on natural resource use, it is important to assess what the appropriate discount rate is for investment decisions which involve the use of natural resources, if indeed, in terms of public project decisions, discounting should be used at all.

The present study involves decision making by upland farmers in the Philippines. The decisions they make involve private investment in the conservation of natural resources, the degradation of which has wide social impacts. One hypothesis tested in this book is whether the market rate of interest equals the private time preference rates of the farmers. Other hypotheses tested relate to various aspects of the farmers' perceived discount rates when making decisions under differing circumstances.

A major implication for policy development as a result of this work is in the choice of interest rate to be used in the provision of credit to farmers for the purpose of implementing soil conservation techniques. This issue is of importance in the global debate on the sustainable use of natural resources.

Notes

1.   This is, significantly, an approach that has not been tried before. The comparison of observed and perceived data collected from identical samples in intensive surveys should help to bridge the gap in knowledge about theory and reality.

2.   Risk is an important factor in determining PTPR. Philippines upland farmers face a high risk of death from accidents, cyclones, political insurgency, malnutrition and epidemics of disease. They also face a high risk of loss of wealth and income because of the system of forfeit inherent in debt default, and from inclement weather and crop failure.

3.   Floro and Yotopoulos (1991) have gone some way towards improving knowledge about informal credit markets in the Philippines, and in the foreword to their book Joseph Stiglitz draws attention to one of the more bizarre aspects of such markets: 'Different lenders have different information about different borrowers, and have different objectives in making loans. Commercial lenders are thus also concerned about the sales of their goods. *More disturbingly, large farmers may be interested in encouraging indebtedness as a means of acquiring control of more land: for them default may be a favourable outcome*' (p.vii; my emphasis).

      In such cases the goal of default itself, rather than risk of default, drives up the interest rate. According to anecdotal information we received while in the Philippines, default is the goal of some lenders. We were told that often the conditions of repayment of loans were such that it was impossible for borrowers to conform (such as by lenders demanding lump-sum repayments).

4.   Present generations should be grateful that artists, writers, architects and builders through history did not conduct a discounted cash flow analysis before starting work – some of the great cathedrals of the world took several hundred years to complete!

5.   Young (1993, p. 22) when commenting on the large body of literature on this subject and the lack of agreement about how the issue may be resolved states: 'If anyone ever resolves it [the issue] to the satisfaction of the economics profession they will probably receive a Nobel prize for doing so.'

Chapter 6

Preliminary Data Analysis

Introduction

In this project several socio-economic variables are assessed. Each respondent in the study is influenced by a wide range of factors when making day to day decisions about the allocation of resources on the farm, in the household and between families. As mentioned earlier, farming families are subject to many risks, including those concerning health, politics and natural disasters. Even with the assistance of statistical analysis it is often difficult to tell which factor will influence most strongly any decision, since factors may be inter-related. It is difficult to know where to begin in constructing socio-economic profiles, particularly because of linkages between socio-economic criteria. However, income is probably the greatest leveller as it is likely to be related to more of the other factors than alternative criteria would be.

Parilla (1992) has shown that income is related to capital assets, farm size, productivity, borrowing and lending profiles, health and soil conservation practices. It seems a promising place in which to begin a preliminary analysis of data.

Total Income on the Farm

Respondents' incomes were summed for each site to give an estimated total gross sample income. The totals between study sites showed large variations, with the greatest income accruing to Canquiason and the smallest accruing to Tabing. The estimated total sample gross incomes per study site are shown in Table 6.1. A summary of mean and median gross income at each site is shown in Table 6.2.

In this project income distribution is very important because it is posited that time preference rate (TPR) is a function of income. Thus if income varies we expect discount rates to vary and, as explored later in this study, this may have an impact on soil conservation adoption. It has also been suggested by at least one researcher that income inequality itself is a cause of environmental degradation (Boyce, 1994).

Table 6.1 Estimated total gross sample incomes at the study sites

Site number	Site name	Total gross sample income (pesos)
1	San Vicente	750,630
2	Pomponan	446,696
3	Canquiason	1,181,458
4	Tabing	262,260

Table 6.2 Mean and median gross incomes at each site

Site number	Mean gross income (pesos)	Standard deviation	Standard error	Median gross Income (pesos)
1	18,766	11,362	3,521	15,021
2	11,167	7,731	2,395	9,396
3	30,291	118,371	18,954	8,349
4	6,875	2,593	420	6,794

To continue the equality-of-income comparisons, within sites, the share of total income among respondents at each site was calculated on a cumulative basis with percentage of total income calculated for percentage of income recipients. This measure can then tell us at a glance how much of the sample village income accrues to what proportion of the villagers.

Lorenz curves were constructed from the cumulative income data (see Todaro, 1977, pp. 110-102). The Lorenz curve is designed to demonstrate the level of income equality or inequality at the study sites. Thus, the percentage of cumulative income, to 100 %, is shown on the vertical axis, and the percentage of income recipients, to 100 %, is shown on the horizontal axis. If income distribution among respondents were equal, 10 %

of the population would have 10 % of the income, 50 % of the population would have 50 % of the income, 90 % of the population would have 90 % of the income, and so on. To show the points of an equal distribution of income a diagonal line of equality is drawn across the graph from 0 % to 100 %. When the Lorenz curve is plotted, its divergence from the line of equality demonstrates a divergence from equality of income distribution. The greater the curve away from the line of equality, the greater the income inequality among respondents at the study site. Lorenz curves for each of the study sites are shown in Figures 6.1 to 6.4.

An additional measure of income inequality is indicated by the income inequality ratio, whereby the ratio of the income earned by the bottom 40 % of earners in the population is compared with the income earned by the top 20 % of earners in the population.

This gives a good indication of the divergence in income between the poorest and richest in a sample, population, or country. The smaller the ratio, the greater the inequality of income.

Income Inequality at the Study Sites

It can be seen that there are huge variations in total income between the study sites, with the total income at the poorest site, Tabing, being only 22 % of the total income at the wealthiest site, Canquiason. However, the poorest respondent at Canquiason is poorer than the poorest respondent at Tabing, much of the variation in wealth between the sites being explained by the fact that the wealthiest respondent at Canquaison earns 63 % of the total income in the sample at that site. However, although this extremely wealthy individual earns a large proportion of the Canquaison sample income, without that person the total income at Canquiason is still nearly double that of Tabing, although the bottom 30 % of both samples earn a similar total income. The median income at site 1 is more than double that at site 4, with median incomes at sites 2 and 3 falling between.

If we look at the Lorenz curves we can see that Tabing has the most equal distribution of income among respondents while Canquiason has the least equal. This does not appear to be directly related to the total wealth of the site however, because San Vicente has the second most equal income distribution and is the second-wealthiest site, while Pomponan has the the third most equal income distribution, yet it is the second-poorest site, in terms of mean income.

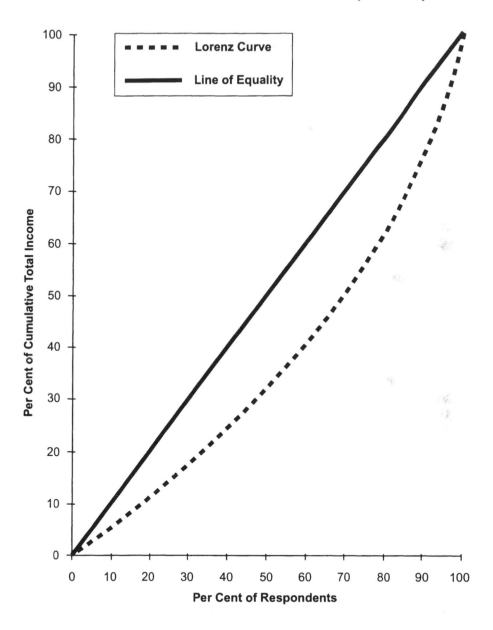

Figure 6.1 – Lorenz curve, Site 1, San Vicente

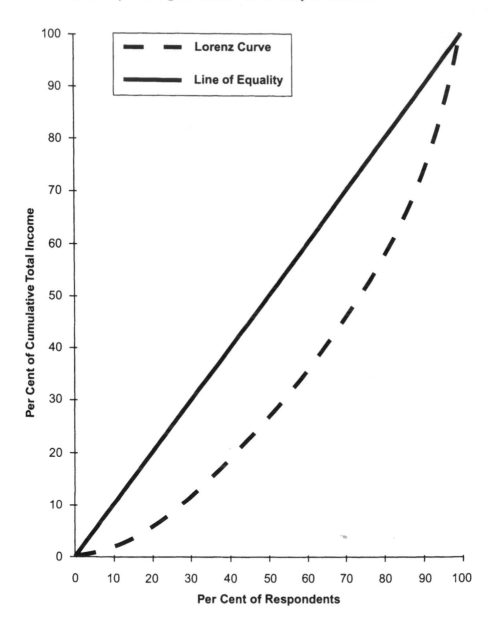

Figure 6.2 -- Lorenz curve, Site 2, Pomponan

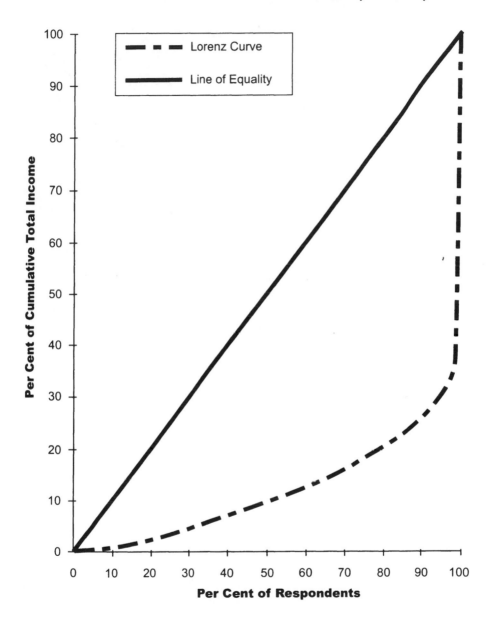

Figure 6.3 – Lorenz curve, Site 3, Canquiason

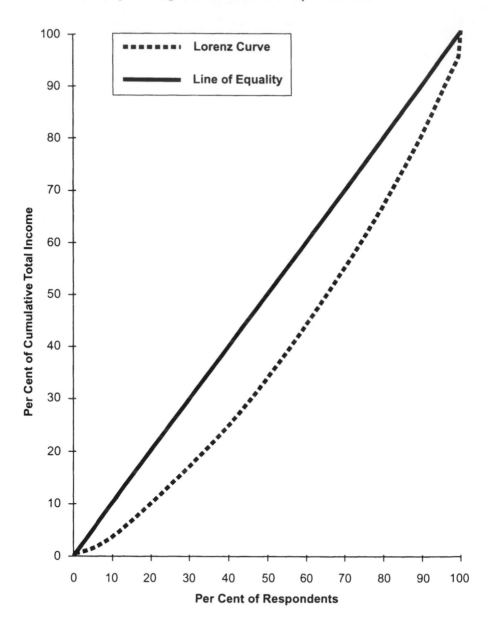

Figure 6.4 -- Lorenz curve, Site 4, Tabing

However, if we assess it in terms of median income, Pomponan is the second-wealthiest site, with a median income of 9,396 Pesos compared with Canquiason's 8,349 Pesos. This again illustrates the skewed income distribution at some of the sites. The income inequality ratios and the site wealth positions are shown in Table 6.3.

Table 6.3 Site wealth positions and inequality ratios

Site number	Site name	Wealth position (medians)	Wealth position (means)	Income Inequality Ratio	Equality position
1	San Vicente	1	2	0.64	2
2	Pomponan	2	3	0.45	3
3	Canquiason	3	1	0.09	4
4	Tabing	4	4	0.78	1
All sites	--	--	--	0.29	--

The distribution of wealth at Canquiason is unequal in the extreme, while at Tabing, although respondents are all poor, they are more equal to each other than those at other sites. The simple inequality ratios shown above are supported by a more complex, but widely used aggregate inequality measure, calculated from the Lorenz curve and known as the Gini Coefficient.

The Gini Coefficient is calculated by estimating the area between the line of equality and the Lorenz curve and by dividing it by the total area of the graph under the line of equality. A Gini Coefficient of zero indicates perfect equality, while a Gini Coefficient of one indicates perfect inequality. A highly unequal distribution of income is indicated by a Gini Coefficient between 0.5 and 0.7 and a relatively equal distribution of income is indicated by a Gini Coefficient of 0.2 to 0.35. Estimated Gini Coefficients for the study sites are shown in Table 6.4.

Table 6.4 Gini coefficients for the study sites

Site Number	Site Name	Gini Coefficient	Equality Rating
1	San Vicente	0.20	2
2	Pomponan	0.26	3
3	Canquaison	0.63	4
4	Tabing	0.11	1
All sites	--	0.36	--

It can be seen that Canquiason's Gini Coefficient demonstrates a highly unequal income distribution while the other sites show a relatively equal income distribution, according to Gini criteria, with Tabing having the greatest equality of income. It is interesting to note that Tabing is the study site which has been the subject of a relatively successful agrarian reform program, where virtually all the respondents are amortising owners; ex-tenant farmers of a former hacienda.

Overall Equality

In addition to estimating income equality at each study site, and assessing within- and between-site differences (which are significant), income inequality within the whole sample was estimated. The same procedure was followed as for each site, with total cumulative income being estimated for the whole sample. A Lorenz curve was plotted, and an inequality ratio and a Gini Coefficient were estimated for the whole sample (all sites) as shown in tables 6.3 and 6.4.

The percentage share of income is shown in Table 6.5. It can be seen that the income distribution for Canquiason has a strong effect, for the overall income distribution, and Gini Coefficient, are pushed towards the highly inequitable range.

Table 6.6 shows the percentage share of the total sample income held by 50 %, 90 % and the highest-earning 20 % of the population.

As expected from the previous results, the order of income inequality amongst sites is unchanged and the share of the income accrued by half the population gives a good indication of relative equality according to the estimated ratios. Aggregated incomes give an idea of the different environments of each village. These environments affect the respondents' private and social time preference rates.

Table 6.5   Average estimated income share - all sites

% Of income recipients	% Share of income
10	2.5
20	6.1
30	10.2
40	15.8
50	21.0
60	27.6
70	35.0
80	42.0
90	55.0
100	100.0

Table 6.6 Percentage share of total income

Site number	Site name	50% of population	90% of population	Top 20% of population
1	San Vicente	31	79	38
2	Pomponan	27	74	43
3	Canquiason	10	28	78
4	Tabing	35	82	32
All sites	--	21	55	53

Tenure, and Borrowing and Lending Activities

The preceding discussion on within-site and between-site differences in income and income equality has established that there is a marked variation in the conditions operating at each site, and that none of the samples is homogenous in this respect. One of the reasons for pursuing this analysis is to assess the variations in the interest paid at the different sites, and to examine them in the socio-economic context.

A summary of the predominant tenure status of farmers at each site is presented in Table 6.7. In some instances farmers have more than one holding (sometimes as many as four or more), and the tenure status of each holding may be different. Predominant tenure status refers to the tenure status of more than 50 % of any farmer's holdings at the site.

It can be seen that predominant tenure status varies markedly from site to site, with only one site exhibiting more than two of the six tenure categories identified in the island. That site, site 1, displays three categories, with 70 % of all landholdings in the sample being held under a share-farming tenancy. The other categories at site 1 are outright owner (22.5 %) and mortagee (7.5 %).

Site 2 and site 3 both have only outright owners or tenants in the predominant tenure status categories. However, at site 2 most are outright owners (65 %) while at site 3 most are tenants (77 %). Site 4 is particularly interesting because it alone has farmers in the amortising owners category, and 68.4 % of the sample is in this group, the rest being tenants. Site 4, Tabing, as mentioned earlier, has been the subject of an agrarian reform program and the amortising owners are the ex-tenants of the hacienda. The sample at site 4 has the lowest total income and the highest equality of all the site samples.

Tenants represent the largest group overall, with 53.5 % of all respondents from all sites being predominantly tenants.

A summary of borrowing activities at each of the study sites is presented in Table 6.8. The information is presented in terms of debt status, tenure status and income hierarchy. Again there is marked variation at all sites, with each site having very different profiles in most respects. In (1) in Table 6.8 the debt/income percentage is shown as a function of mean debt against mean income per respondent at each site.

Borrowers at site 4 (the poorest site) owe only 9 % of their income, while at site 3, on average, respondents owe 21 % of their income. In (2) the total site sample debt is shown as a percentage of total site sample income. These percentages are lower overall than in (1) because they include non-borrowers' income. Again site 4 has the lowest debt/income ratio (5 %) while site 1 has the highest.

Interestingly, as Table 6.7 shows, site 2 has by far the highest number of outright owners while site 4 has none (see (3)). However, site 2 is the second-poorest site, in terms of site income, while site 4 is the poorest. It is clear that income is not the direct determinant of debt in terms of whole-of-site analysis.

Table 6.7 Predominant tenure status

Site no.	Sampl size	Full Own'r (%)	Mort-gagee (%)	Family share (%)	Tn'nt (%)	Inheri-tor (%)	Amort-ising Owner (%)
1	40	22.5	7.5	0	70.0	0	0
2	40	65.0	0	0	35.0	0	0
3	39	23.0	0	0	77.0	0	0
4	38	0	0	0	31.5	0	68.5
All sites	157	28.0	1.9	0	53.5	0	16.5

Table 6.8 Summary of borrowing activities

	Site 1	Site 2	Site 3	Site 4
(1) Debt as % of mean income	18	11	21	9
(2) Site debt/income (%)	14	7	3	5
(3) % tenants in sample	70	35	77	31
(4) % tenants in borrowers	93	23	89	37
(5) % borrowers in sample	67	67	49	64
(6) % Low 40% income borrowing	62	75	31	50
(7) % high 20% income borrowing	50	87	25	50
(8) % mid 40% income borrowing	81	50	80	87

Sites 1 and 3 have the highest number of tenants among borrowers. This is unsurprising as these sites have the highest number of tenants in the sample overall. However, site 1 has fewer tenants than site 3, but more tenants among borrowers (see (3) and (4)).

In (6), (7) and (8) in Table 6.8, borrowing activities are presented according to the income groups used to determine the inequality ratios presented in Table 6.3. There is some consistency in these results in that at three of the four sites 80 % or more of the middle class (the middle 40 % of income earners) are borrowing. However at site 2, which has the greatest number of borrowers overall, fewer middle-income earners are borrowing (50 %), compared with 75 % of the lowest-income earners and 87 % of the higher-income earners. Neither is there a clear pattern at the other sites between income group and borrowing for the higher and lower income groups. The relationship between willingness or ability to borrow, and income group, will be examined more closely when the discount rate survey is analysed in Chapter 7.

Chapter 7

Observed Interest Rates and Perceived Discount Rates among Upland Farmers

Introduction

Most economists would accept that, given a free capital market, what farmers are actually paying for interest on debt equates with their private time preference rates. This in turn premises that the market price for credit at the study sites is reflecting an open system of supply and demand, and that the farmers have freedom of choice both in their decisions to borrow and in their credit supplier. These issues are assessed, and the economic theory on which they are based is tested, as a result of the data analysis of the interest and discount rate surveys presented in this chapter.

As described earlier, in 1993 a survey of the farmers' perceived discount rates was administered at the study sites, using identical respondents as for the earlier socio-economic surveys. This discount-rate survey comprised a range of questions designed to assess the private time preference rates (PTPRs) of farmers in the sample. These questions were presented in both willingness-to-pay (WTP) and standard-bidding formats. The results of the observed interest rate surveys and the discount-rate survey were then compared and analysed, to see whether basic assumptions about interest rates and discount rates in economic theory could be applied validly to these farming communities.

In Chapter 6 some socio-economic characteristics of respondents at the study sites were examined (socio-economic site profiles are summarised in Table 7.1). There were large variations in income and debt status between respondents at each site, and between sites. It was also noted that the observed interest rates on borrowings were very high at all sites, particularly at site 4. In a functioning capital market, standard economic theory predicts that these observed interest rates would equate with the PTPRs of the respondents, demonstrating willingness to pay for credit, and hence would reflect their preference for present consumption over future consumption.

Table 7.1 Site profile summary

Criterion	Site 1	Site 2	Site 3	Site 4
Population	701	3,105	1,500	1,824
Tenure	Tenant (75%)	Owner (56%)	Tenant (79%)	Amortiser (71%)
House size	5.5	5.0	4.5	6.0
Land size	1.7 ha	1.8 ha	2.0 ha	1.1 ha
Transport	Yes	Yes	No	Yes
Electricity	Yes	Yes	No	Yes
Water	Spring	Spring, piped	Spring	Spring
School	Primary	Primary, high	Primary	Primary
Religion	Catholic 7th Day	Catholic 7th Day	Catholic	Catholic
Nearest town	3 km	6.5 km	11.5 km	
Soil erosion	Mild	Severe	Severe	Severe
Main crop	Corn	Coconut	Rice	Corn, Rice
Health centre	Yes	Yes	Yes	Yes
Bank/credit co-op	Both	Both	Neither	Neither
Mean interest % (median)	110 (118)	176 (116)	178 (143)	409 (436)
Mean age h'hold head	46.7 yrs	51.0 yrs	45.4 yrs	40.0 yrs

Table 7.1 (Continued)

Criterion	Site 1	Site 2	Site 3	Site 4
Mean yrs educat. H'hold head	4.3	5.0	4.2	4.3
Income equality ratio	0.64 (fair)	0.45 (poor)	0.09 (v bad)	0.78 (good)
Gini coefficient	0.20	0.26	0.63	0.11
% practising soil conservation	43	50	12	42
Median income (pesos)	15,021	9,396	8,349	6,794

This theory would in turn help to explain some of the respondents' decision-making processes – for example, why farmers choose to conduct certain on-farm activities, while forgoing others which may have present costs and future benefits (such as soil conservation).

A series of hypotheses to be tested during the course of data analysis in this chapter and in Chapter 8 are presented below. Some of the hypotheses have been proposed because of expectations arising from standard economic theory (see Chapter 5), and some have been proposed because of indications in the survey data.

The seven hypotheses developed on the basis of theory and experience are as follows:

- Hypothesis one: *there is no significant difference between the mean interest rate paid on debt and the time preference rate of individuals.*
- Hypothesis two: *there is no significant difference between the mean perceived discount rates inferred from willingness-to-pay (WTP) questions and those inferred from standard bidding questions for PTPR (for 1000 pesos over one year).*
- Hypothesis three: *there is no significant difference between the mean interest paid on cash loans and the mean interest paid on in-kind loans.*
- Hypothesis four: *There is no significant difference between the mean WTP discount rate when the individual is told he or she will avert a*

severe loss if the money is borrowed, and the mean WTP discount rate when the individual is told he or she will make a large gain if the money is borrowed.

- Hypothesis five: *there is no significant difference between the perceived mean PTPR discount rate inferred from the standard bidding question and the perceived mean WTP discount rate where borrowing is to avert a severe financial loss.*
- Hypothesis six: *there is no significant difference between the mean discount rates of respondents who adopt soil conservation practices and the mean discount rates those who do not.*
- Hypothesis seven: *there is no significant difference between mean observed interest rates paid by adopters and those paid by non-adopters of soil conservation.*

Preliminary Comparison of Observed Interest Rates and Inferred PTPRs

In a preliminary comparison of results from the different survey approaches, observed mean and weighted-mean interest rates from the earlier surveys are compared with private time preference rates inferred from the WTP and standard-bidding questions of the later survey, for the sample at each site. These are presented in Table 7.2.

Hypotheses one and two are examined in the preliminary assessment to put the analysis of borrowing behaviour and observed interest rates into perspective, and to provide a preview of the discount rate survey into the context of the credit data and the socio-economic profiles of the study sites.

Comparison of weighted with unweighted mean observed interest rates shows that at all sites except site 4 the difference is not large, indicating that interest charged does not vary a great deal according to the size of the loan.[1] At site 4 the weighted mean is a lot higher, indicating that higher interest is charged on larger loans there.[2]

Henceforth all mean interest rates quoted in the text or in tables refer to weighted means, except in some categories where the number of loans in a category is too small to weight, or where stated otherwise.

Table 7.2 Observed mean interest rates and perceived discount rates, and mean income, at all sites

Criterion	Site 1	Site 2	Site 3	Site 4
(1) Observed interest rate %:				
mean (median)	118 (120)	184 (169)	176 (153)	336 (229)
wtd mean (median)	110 (118)	176 (116)	178 (143)	409 (436)
(2) Perceived discount rate %: WTP PTPR (1000p over 1 year; Q1)				
Borrowers	58	50	46	43
Non-borrowers	70	51	50	37
All	62	50	47	41
(3) Perceived discount rate %: Std Bid. PTPR (1000p over 1 year; Q14)				
Borrowers	49	44	40	46
Non-borrowers	51	44	41	44
All	50	44	41	46
(4) Mean (med.) farmer income*	18,765 (15,021)	111,67 (9,396)	30,291 (8,349)	6,901 (6,794)
(5) Mean income*, lower 50%	11,635	6,054	6,058	4,831

*In pesos

The credit survey results results show some degree of variation in interest rates paid by farmers at the sites. The massive mean interest rate at site 4 (409%) may offer some insight into why there is such a low incidence of borrowing at this site compared with the others. The other sites have lower rates which are less variable (110-178%). However, by Western standards these rates of interest are extremely high.[3]

The results shown under (2) in Table 7.2 are PTPRs drawn from the discount rate survey question in which all respondents were asked to indicate how much interest they would be *willing to pay* on a loan of 1000 pesos for a year. The mean PTPRs thus inferred for all respondents varied from 41% (site 4) to 62% (site 1). Except at site 4, non-borrowers indicated a higher willingness-to-pay PTPR than the borrowers at each site.

The PTPRs inferred from the standard-bidding question are shown under (3) in Table 7.2. Respondents were asked to indicate how much interest they would want *to be paid* for deferring receipt of 1000 pesos for one year. The mean results were remarkably consistent over all sites for borrowers and non-borrowers, with a total range of 41% to 50%, and a range in the mean of all respondents of only 9.0%. This standard-bidding PTPR was generally less than the WTP PTPR (with which in theory it should equate) except at site 4, and was much lower than the observed mean interest rate being paid by borrowers. There does not appear to be direct correlation between mean income and mean standard-bidding PTPR, mean WTP PTPR and mean observed interest rate.

Preliminary Assessment of Interest Rates and Standard-Bidding PTPR

The above results permit testing of hypothesis one, which states that mean observed interest rates will not differ from the mean inferred PTPRs. In this case the mean PTPRs inferred from the standard-bidding question have been used, because the standard-bidding technique is (as its name implies) the commonly accepted technique for assessing PTPRs from a survey. The relevant data were subjected to statistical testing using a Z test; the results are presented in table H1. Table H1 shows that there is a significant difference between interest paid on debt and PTPR at all sites. Therefore the null hypothesis is rejected and the alternative hypothesis, that there *is* a significant difference between interest rates and PTPR, is accepted. In other words, in this instance interest rate paid is not a suitable shadow for discount rate.

Table H1 Summary of statistical testing for hypothesis one

Criterion	Site 1	Site 2	Site 3	Site 4
Mean obs. interest Rate (%)	110	176	178	409
s.d.	35.6	94.7	88.4	240.6
N	26	26	18	24
Mean std. bidding PTPR (%)	50	44	41	46
s.d.	18	17	17	18
N	35	39	37	36
Z	7.9*	7.0*	6.5*	7.4*

*Significant at least at the 95% level

Preliminary Assessment of Standard-Bidding PTPR and WTP PTPR

Hypothesis two states that PTPRs inferred from standard-bidding questions will not differ from those inferred from WTP questions, for PTPR for 1000 pesos for one year. The relevant data and the results of a Z-test of that data are presented in table H2.

The results of the Z-test suggest that hypothesis two ought to be rejected for sites 1 and 2, but accepted for sites 3 and 4. That is, there is a significant difference between standard-bidding PTPR and WTP PTPR at sites 1 and 2, but there is no significant difference at sites 3 and 4.

Table H2 Summary of statistical testing for hypothesis two

Criterion	Site 1	Site 2	Site 3	Site 4
Mean std. Bidding PTPR (%)	50	44	41	46
s.d.	18	17	17	18
N	35	39	37	36
Mean WTP PTPR (%)	62	50	47	41
s.d.	26	15	37	21
N	35	39	37	36
Z	2.2*	1.66*	0.9	1.08

* Significant at least at the 95% level

Regression Analyses

Regression analyses were run to test hypotheses one and two when data for all sites were aggregated. It was found that the results of hypothesis testing for the individual sites were supported generally by the regression results. The regressions were ordinary least-squares regressions taking the form of

interest on debt = α + β Std-bidding PTPR, for hypothesis one,

and,

WTP PTPR = α + β Std-bidding PTPR, for hypothesis two.

For both hypotheses, $t_{crit} = t_{85,0.95} = 1.96$

For hypothesis one, $t_{crit} = -0.7950 < 1.96$

and for hypothesis two $t_{crit} = 2.151 > 1.96$

For hypothesis one there is *no* statistically significant relationship between interest paid on debt and standard-bidding PTPR and the null hypothesis is *rejected*, which is in keeping with the site-by-site analysis.

For hypothesis two there *is* a statistically significant relationship between WTP PTPR and standard-bidding PTPR and the null hypothesis is *accepted*.

Thus for all results when aggregated there is no difference between WTP PTPR and standard-bidding PTPR responses. We can accept that for the overall data it reasonable to use both WTP and standard-bidding methods for eliciting farmers private time preference rates and hence for perceiving their discount rates.[4]

Regression analyses were not run for the other hypotheses because the aim of the regressions was to test whether generally interest on debt equalled PTPR, and whether generally WTP PTPR equalled the standard bidding PTPR. For the other hypotheses the site-specific differences are important to the analysis and aggregating the data for a regression would have obfuscated this issue.

PTPR, Interest and Circumstance

Each individual is assumed to have a discount rate which he or she applies (usually unconsciously) when making decisions. Some economists are beginning to question whether every person applies the same rate of discount to all situations. As Knetsch (1993, p. 2) states:

> ... it increasingly seems that the use of a single discount rate to reflect peoples' time preference with respect to all dimensions of future outcomes, may not accurately reflect peoples' actual preferred intertemporal choices. A good bit of anecdotal evidence, and increasingly the findings of controlled studies, suggest that individuals have widely different time preferences that vary depending on the characteristics of the particular case.

In this chapter the assumption that an individual's private time preference rate remains constant is examined. Observed interest rates are compared with the farmers' PTPRs elicited in response to the survey questions assessing willingness to pay for credit in a range of circumstances.

Observed Interest Rates

All of the respondents at each site were asked a number of questions relating to their borrowing activities. These questions included whether the loan was formal or informal, cash or kind, what the source of the loan was (usurer, bank, co-operative, or relative/neighbour), the amount of the loan and its duration, the method and amount of repayment, and the purpose of the loan. It was then possible to calculate the effective observed interest rate of the loan, and to make comparisons between purposes, sources etc.

It turned out to have been a mistake to couple relative and neighbour as a source of loan category, because it became clear, when individual loans were examined, that some neighbours were acting as usurers and charging very high interest rates.

Conversely some of the loans in this category had zero interest rates and were probably from wealthier altruistic relatives. Nevertheless, there was a difference at each site between loans in the usurer category and those from neighbours/relatives.

All sites had a large proportion of borrowers in their samples, the highest being at San Vicente and Pomponan (both 67 %) and the lowest being at Canquiason (40 %). The two sites with the highest borrowing rates both had access to banks and co-operatives, while the other sites only had access to usurers or neighbours/relatives.

The breakdown of purpose of borrowings is shown in Table 7.3, with a summary of mean interest rates paid at each site for farm input and food loans. As we would have expected, the greatest proportion of borrowings at all sites was for farm inputs (planting materials, fertilisers, chemicals, animals, cultivation and labour) with site 4, Tabing, having the highest rate (54 %) of all borrowings for farm inputs, and site 2, Pomponan, having the lowest rate at 37 % of all borrowings for farm inputs.

However, the second highest rate of borrowing (15 % to 42 %) was for food, which is probably indicative of the poverty at these sites. The site with the lowest incidence of borrowing for food, site 3, Canquiason, had the highest incidence of borrowing for medicine. Given these borrowing patterns, where a large number of loans are for food (or medicine), it is not really surprising that the rate of adoption of soil conservation technologies is low.

Farmers who borrow money so that they can eat, are not likely to be spending money on farm activities for which they can expect no return for a number of years.

Table 7.3 Purpose of loans

Loan purpose	Site 1	Site 2	Site 3	Site 4
% Farm inputs	46.7	36.9	47.5	53.6
% Food	41.5	41.5	15.0	32.6
% Burials	0.0	0.0	2.5	3.6
% House repairs and establish.	2.6	6.1	0.0	1.8
% Land Purchase	1.3	0.0	2.5	0.0
% Previous debt	1.3	1.5	2.5	1.8
% Medicine	2.6	4.6	12.5	3.6
% Travel	3.9	3.1	10.0	1.8
% Children's education	0.0	6.2	7.5	1.8

Mean interest rates on loans for food and farm inputs are shown in Table 7.4, along with the mean interest rates charged for loans, according to source. Site 4, Tabing, has by far the highest interest rates of any of the sites, with an unweighted mean of 411 % for loans for farm inputs and 318 % per annum on loans for food. Site 1, San Vicente has the lowest interest rates with an unweighted mean of 118 % for farm input loans, and 123 % per annum for loans for food.

When looking at the mean interest rates according to the source of the loan, we see, as expected, that rates from banks and co-operatives are generally lower than for those from other sources. However, by Western standards they are still extremely high, ranging from 27 % p.a. to 111 % p.a. There is some variation to these trends, for at site 1 the mean neighbour/relative rate is lower than the mean co-op rate; perhaps there are more relatives than usurers in this category at this site. At site 3 (which has no access to banks or co-operatives) the relative/neighbour rate was lower than the bank mean rate at site 2 and the mean co-op rate at site 1.

As indicated above, the profiles of the four sites are quite different. There is variation in the mean income of respondents and in overall equality. The mean income at site 4 is very much lower than for the other sites, even when the top-earning 50 % of the sample is excluded. Some of this can probably be attributed to the differences in tenure status at the various sites, and by the difference in mean farm size.

Table 7.4 Summary of mean interest rates, according to purpose and source of loan

	Site 1	Site 2	Site 3	Site 4
Mean interest rates according to purpose (% p.a.)				
Farm inputs	118	220	196	411
Food	123	152	163	318
Mean interest rates according to source of loan (% p.a.)				
Usurer	136	255	186	480
Bank	27	108	--	--
Co-op	111	72	--	--
Neighb./Relative	99	179	95	286
Borrowings of soil conservation practitioners compared with total borrowings				
Site borrowing rate %	67	67	48	60
% of soil cons. practs. Borrowing	56	67	40	61
Mean cons. Practs Loan value as % of mean site loan value	24	40	25	37

With respect to soil conservation adoption, at sites 2 and 4 the rate of borrowing among those practising soil conservation is equivalent to the overall borrowing rate among respondents at the sites, but at sites 1 and 3 fewer of those practising soil conservation are borrowing and the mean value of borrowings at all sites is lower among those practising soil conservation.

In Table 7.5 a summary of means is presented. The unweighted mean interest rates are compared with the weighted mean interest rates, where means were weighted according to the amount of money being borrowed at a particular interest rate. At all sites, except site 4, the weighted mean was lower than the unweighted mean. At site 4 the weighted mean was 18 % higher than the unweighted mean. This difference may be explained in part by the lack of formal borrowing facilities at the site, and by the overall poverty of the villagers (who according to their inequality ratios are more equal in their poverty than those at other sites).

Table 7.5 Loan interest, loan value and farmer income:
summary of means and medians

Site number	Unw'ted mean (median) interest rate (%)	Weighted mean (median) interest rate (%)	Mean total value of loans per Resp'dnt (pesos)	Mean (median) income (pesos)	Mean income lowest 50% of sample (pesos)
1	118 (120)	110 (118)	3118	18765 (15021)	11,635
2	184 (169)	176 (116)	2914	11,167 (9396)	5286
3	176 (153)	178 (143)	2166	30,291 (8349)	6058
4	336 (229)	409 (436)	562	6902 (6794)	4831

This equality in poverty at site 4 means that the villagers may be unable to turn to wealthier friends and relatives for cheaper loans and are forced to borrow from usurers. These usurers charge very high rates of interest, as reflected in the inflated mean interest rates at the site. As would be expected, the villagers who are charged these high interest rates borrow much less than other villagers. This lower borrowing rate is reflected in the mean loan amount per borrower.

At site 1 where the weighted mean interest rates are lowest (110 %) the mean loan value per borrower is 3118 pesos. At site 4 where the weighted mean interest rate is 409 % per annum, the mean loan value per borrower is only 562 pesos. When the mean income of only the lowest 50 % of income earners is considered it can be seen that the highest mean is for site 1 (11635 pesos per year) and the lowest is at site 4 (4831 pesos per year). The site incomes would also have some effect on borrowing activities.

It is interesting to assess who is borrowing at the different sites, and what the debt-to-income ratios are. If we look at Table 7.6 we can see that the poorest site, site 4, has the lowest percentage of debt as a function of mean income (9 %), while the second-poorest site, site 2, has the highest percentage of debt as a function of mean income (39 %).

Table 7.6 Summary of borrowing activities

	Site 1	Site 2	Site 3	Site 4
Debt as % mean income	18	25	21	9
Site debt/income %	14	26	3	5
% tenants in sample	70	35	77	31
% tenants in borrowers	93	23	89	37
% borrowers in sample	67	67	49	64
% low 40% borrowing	62	81	31	50
% high 20% borrowing	50	75	25	50
% mid 40% borrowing	81	50	80	87

At all sites, excluding site 2, the middle 40 % of income earners have the highest borrowing rate. However, at site 2 the bottom 40 % of income earners have the highest borrowing rate (81 %), closely followed by the top 20 % (75 %).[5]

A summary of overall credit received is presented in Table 7.7. The outliers have been excluded because of the possible margin of error in recorded values in the original data set. Outliers are those whose loan interest rates are estimated to be less than 1 % or more than 1000 %. In some cases there were insufficient data to estimate an interest rate value.

Table 7.7 Summary of interest rates on credit received

	Site 1	Site 2	Site 3	Site 4
Mean (%)	110	176	178	409
Median (%)	118	116	143	436
s.d.	35.6	94.7	88.4	240.6
No. of loans	73	58	40	48
No. respondents	40	40	39	38
% outlying loans	5	11	0	16
No. borrowers included	26	26	18	24

In Table 7.7, the standard deviations for the interest rates are very high, showing the large amount of variation present in the data. The number of loans recorded, and the outliers excluded from the overall calculation, are shown. In Tables 7.8 and 7.9, the information is disaggregated into cash loans and in-kind loans, and demonstrates a high degree of variability according to the cash or kind nature of the loan.

Table 7.8 Summary of interest rates on cash loans

	Site 1	Site 2	Site 3	Site 4
Mean interest rate (%)	117	172	128	172
Median int. rate (%)	119	100	143	100
s.d.	63	185	40	146
Min. interest rate (%)	23	5	70	100
Max. interest rate (%)	524	750	250	577
Number of cash loans	66	25	26	21
% cash loans	90	45	65	44
Standard error (95% confidence)	± 19.5	± 57	± 12	± 46

Table 7.9 Summary of interest rates paid on in-kind loans

	Site 1	Site 2	Site 3	Site 4
Mean interest rate (%)	120	194	267	463
Med. interest rate (%)	109	200	284	430
s.d.	14	100	68	194
Min. interest rate (%)	108	18	114	90
Max. interest rate (%)	153	400	450	793
No. in-kind loans	7	33	14	27
% Loans in-kind	10	55	35	56
Standard error	± 4.0	± 31	± 21	± 62

At all sites the mean for cash loans is lower than for in-kind loans, and that at all sites but site 1, where it is very close, the median for cash loans is much lower than for in-kind loans. At site 1 the incidence of in-kind loans is low (10 %) compared with other sites. At sites where formal credit is available (sites 1 and 2) the minimum value for cash loans is low compared with sites where no formal credit is available (sites 3 and 4). At all sites the maximum interest rate paid on credit is extemely high.

The Perceived Discount Rate Survey

Interest and Discount Rates

The perceived discount rate survey was administered in 1993 to a sample identical to that administered the borrowing (revealed interest rate) survey. One of the purposes of this later survey was to compare the farmers' perceptions of the interest they were prepared to pay with what they actually paid. There were 41 questions in the survey, which was also designed to test the validity of economic theory by enabling assessment of differences in perceived interest rates when the amount, duration and nature (cash or kind) of the loan varied, and to see if farmers exhibited different discount rates according to whether they were borrowing or lending.

There were two questions in the survey designed to assess the private time preference rate of farmers, according to a standard bidding technique, for deferring receipt of a sum of money for one year (questions 14 and 30). The first of these questions was for a gift of 1000 pesos, and the second was for a gift of 200 pesos. Respondents were asked to bid in amounts that were effectively 5 % at a time up to 100 % in order to nominate the amount of money they would require to compensate them for deferring receipt of the gift. The last two choices were open-ended in order to enable respondents to nominate an amount equalling less than 5 % or more than 100 % (although the actual interest rate was not made explicit in any of the choices). None of the respondents selected less than 5 % or more than 100 % in this category of question.

The results to these questions were remarkably consistent, with the mean private time preference rate (PTPR) for each question being the same for both initial sums at all sites except site 1, where the mean rate of interest required to defer receipt of the 1000 peso gift was 3 % lower than for the 200 peso gift (see table 7.10). The estimated PTPR ranged from 41 % at site 3 to 53 % at site 1. There was some degree of variation between respondents at each site, but generally speaking the higher income earners appeared to display higher discount rates.

The next category of question to be considered was that relating to willingness to pay for a loan, starting with 1000 pesos (Q. 1) to enable direct comparison with the PTPR question. In theory, this result should be directly compatible with the PTPR discount rate, as the amount of interest an individual is prepared to pay theoretically correlates directly with their private time preference rate. However, at all sites but site 4, the mean willingness to pay (WTP) for 1000 pesos for one year was between 6 % and 9% higher than the perceived PTPR. At site 4, the mean WTP was 5 % lower than the mean perceived PTPR. There could be several reasons for this, one being that at site 4 the respondents do not appear to like borrowing money, as discussed earlier, and borrow small amounts compared with those at the other sites. Mean WTP PTPR rates are shown in Table 7.10, along with the mean standard-bidding PTPRs, the weighted mean revealed interest rates paid on loans, and mean site income (100 % of respondents and lower 90 % of respondents).

Table 7.10　Mean revealed and perceived interest rates (% per annum)

Site nunber	Mean income (pesos)	Mean income, low 90% (pesos)	Mean median interest rate	WTP PTPR 1000 pesos ± s.e.95%	Std-bid. PTPR 1000 pesos	Std-bid. PTPR 200 pesos
1	18765	14824	110	62 ± 8.6	50 ± 6.0	53 ± 7.6
			118			
2	9788	6852	176	50 ± 4.7	44 ± 5.4	44 ± 5.4
			116			
3	30291	8483	178	47 ± 12	41 ± 5.5	41 ± 6.1
			143			
4	6901	5659	409	41 ± 6.6	46 ± 5.9	45 ± 5.9
			240			

As shown by earlier hypothesis testing, the WTP-PTPR and standard-bidding PTPR results are in the same order of magnitude, and show no significant difference at two sites. This is not the case when the revealed mean interest rates are compared with WTP-PTPR and standard bidding

PTPR. In theory, these results should also be closely correlated with each other, but hypothesis testing has shown that there is no statistically significant relationship between them. Observed interest rates are very much higher than the perceived discount rates, and respondents are paying interest at a rate far higher than they say they would be willing to pay, or than they would demand to be paid to defer receipt of money for one year. It might be that farmers are unable to exercise freedom of choice in their actual borrowing activities. There does appear to be some correlation between mean site income (especially at the poorer 90% level) and perceived discount rates, with site 1 (wealthiest at the poorer 90% profile) having the highest mean WTP-PTPR and standard-bidding PTPR and site 4 (poorest at any profile) having the lowest WTP-PTPR. The lowest standard-bidding PTPR is at the site with the lowest equality ratio, site 3. This is not consistent with the assumption that poorer people have a higher time preference rate, but it is possible that the responses are modifed by the farmers' perceptions of their ability to pay (although the money amounts offered in the questions are not large and there is little variation in response when the amount offered changes from 1000 pesos to 200 pesos).

Varying the Duration and Value of Loans

Once the willingness to pay (WTP) for a standard amount (1000 pesos for one year) had been assessed and compared with the revealed interest rates at each site, and with the results of the conventional PTPR questions, the effect of varying the amount and durations of loans for cash and kind was examined. The durations considered were 5 years, 1 year, 5 months and 1 month. The amount of cash offered for loan was either 1000 pesos or 200 pesos, with each amount being offered for each duration. The same questions were then asked for the in-kind loans for rice or corn (depending on the staple at the study site) for food. One question was posed for an in-kind loan for seed. (This series corresponded to questions 2, 3, 4, 6, 7, 8, 9, 10, 11, 12, 15, 16, 17, 18 and 20 in the survey.) The final survey question (41) was designed to test the response to an offer of a loan for a huge amount of money (in the farmers' terms) of 10,000 pesos for a year (more than the annual income of many respondents) to see if this quantity has any obvious effect on WTP.

Unfortunately, the responses to the five-year questions for kind appear to be unreliable as they seem to have caused confusion among some respondents. It is clear from looking at the data that, while the question was posed to discover how much farmers would pay *per annum*, many respondents appear to have stated how much interest they would pay over the five years. Other respondents have answered on a per annum basis, and because of the range of interest rates that farmers are willing to pay, it is not possible to assess with certainty which response farmers are giving. This confusion only appears to hold for sites 2 and 3, but because of the lack of rigour in the results the responses have been discarded.

Consolation can be taken from assessing data for cash loans, for the five-year duration responses do appear to follow a general trend, showing a slightly lower WTP for cash loans of 1000 pesos than for the other periods.

Results of this analysis are presented in table 7.11 and we can see that overall the greatest differences in WTP occur *between* sites rather than *within* sites, although varying conditions obviously does have an impact on WTP within sites.

However, effects seem quite localised in that preferences seem to follow trends in locality, which supports the notion that many aspects of daily life and decision making stem from social and cultural aspects of daily life. For example, at site 1 the WTP for 10,000 pesos for one year was much lower than for most other WTP responses at that site, while at site 4 the WTP for 10,000 pesos was around the same as for most other responses.

Overall site 4 (the poorest site) had a lower WTP-PTPR than site 1 (the wealthiest site at the lower 90% of earners level). Yet respondents at site 1 are *actually* paying the lowest interest rates while site 4 respondents are *actually* paying by far the highest interest rates. Looking at the overall means for each site we can see that at all sites the mean WTP was lower for kind than for cash, yet if we look at tables 7.8 and 7.9 we can see than observed in-kind interest rates are higher than interest rates on cash loans. The effect of duration on WTP varies between sites. At sites 1, 2 and 3 the mean WTP rises as the duration falls. However, at site 4 the WTP falls as duration falls.

None of the questions specified what would happen if money was not borrowed, and loan purpose was not specified for the cash loans. While all the WTP PTPR responses are in the same order of magnitude, there is some definite variation in the WTP PTPR as the nature and conditions of the loans changes, with the difference between the minima and maxima at each site being as follows: Site 1, 34 %; Site 2, 28 %; Site 3, 27 % and Site 4, 26 %.

The amount of the loan generally had less impact than the duration of the loan. The in-kind loans for each period were designed to have the same value as the cash loans (100 gantas is worth about 1000 pesos and 20 gantas is worth about 200 pesos).

Table 7.11 WTP PTPRs with varying amount, duration and nature of the loan (% per annum; mean (median))

Nature of the loan	Site 1	Site 2	Site 3	Site 4
1000 pesos 5 years	72 (74)	68 (74)	55 (43)	62 (58)
1000 pesos 1 year	64 (60)	50 (60)	47 (30)	41 (36)
1000 pesos 5 months	77 (71)	58 (71)	56 (33)	46 (40)
1000 pesos 1 month	87 (80)	64 (80)	61 (34)	49 (43)
200 pesos 5 years	77 (74)	66 (74)	63 (62)	62 (58)
200 pesos 1 year	64 (60)	50 (60)	43 (40)	40 (36)
200 pesos 5 months	70 (65)	53 (65)	55 (38)	48 (50)
200 pesos 1 month	60 (60)	48 (60)	36 (36)	42 (36)
10,000 pesos, 1 year	53 (60)	51 (60)	42 (40)	42 (48)
100 gantas seed 1 year	50 (50)	50 (49)	50 (50)	50 (50)
100 gantas 1 year	64 (60)	40 (30)	45 (40)	41 (36)
100 gantas 5 months	66 (71)	42 (31)	47 (46)	46 (40)
100 gantas 1 month	65 (60)	42 (48)	48 (42)	36 (36)
20 gantas 5 years	unreliable data	unreliable data	unreliable data	unreliable data
20 gantas 1 year	65 (60)	40 (60)	55 (60)	65 (30)
20 gantas 5 months	60 (65)	41 (65)	41 (31)	41 (31)
20 gantas 1 month	66 (60)	48 (60)	60 (30)	40 (30)
All cash [observed]	69 [117]	56 [172]	51 [128]	48 [172]
All kind [observed]	62 [120]	43 [194]	49 [267]	45 [463]
All 1 year	60	47	47	46
All 5 months	68	48	50	45
All 1 month	69	50	51	42

When we compare the results of the actual interest paid in the observed interest rate survey, and the stated willingness to pay in the perceived discount rate survey, we can see that great caution must be used when applying the assumption that interest rates paid provide a suitable shadow for discount rates. It is clear that there is a huge divergence between observed interest rates and discount rates perceived from the discount rate survey. The WTP for kind is always lower than for cash, yet we know that respondents are paying more for in-kind loans - they just don't want to! We might reasonably expect interest paid on cash to be greater than interest paid on kind if there were a difference, because cash gives rise to a greater range of choice than kind.

Hypothesis three is now tested to see if there is any difference between actual interest paid in cash loans compared with kind loans at the study sites, given that the WTP for kind does appear to be lower overall. Hypothesis three states: there is no significant difference in mean interest paid on cash loans compared with that paid on in-kind loans. The null hypothesis is accepted for sites 1 and 2 and rejected for sites 3 and 4. These site-specific differences may relate to the availability of formal credit, which is not available at sites 3 and 4.

Table H3 Summary of statistical testing for hypothesis three

Criterion	Site 1	Site 2	Site 3	Site 4
Cash loans				
Mean int. rate (%)	117	172	128	172
s.d.	63	185	40	146
N (no. of loans)	66	25	26	21
In-kind loans				
Mean int. rate (%)	120	194	267	463
s.d.	14	100	68	194
N (No. of loans)	7	33	14	27
Z	0.3	0.54	7.3*	5.9*

*Significant at least at the 95 per cent level

Borrowing to Avert Loss or Attract Profit

The last section discussed the farmers' willingness to pay for a loan of 1000 pesos, and their PTPRs for deferring receipt of 1000 pesos. The mean PTPR rates in this category of question were between 41 % and 62 % per annum, much less than the observed interest rates at each site. In the next category of question (questions 21, 23, 24, 25 and 26 in the survey) farmers were asked how much interest they would be prepared to pay for a loan of 1000 pesos over a year, but the circumstances of the loan were specified. For the five questions in this category, borrowing was either to avert a severe loss or to attract a large profit.

The responses to questions 21 and 25, the most general questions, provide data to test null hypothesis four, which states: there is no significant difference between the mean WTP discount rate when individuals are told they will avert severe financial hardship if they borrow money and that when they are told they will make a large profit on all produce if they borrow money.

In the first question farmers were asked how much interest (in pesos) they would be prepared to pay on a loan which would prevent them suffering severe financial hardship in the near future. These results were very interesting in that the mean interest rates at all sites were very much higher than for the straightforward WTP and standard-bidding PTPR questions (112 % to 120 % compared with 41 % to 62 %) and were more closely correlated with the observed interest rates. In addition, the bidding process was the same as for the earlier questions, with two open-ended options to allow for bidding for more or less than the explicit responses (which again rose by an effective 5 % each time). Whereas no farmers had opted for either of the open-ended responses in the earlier questions, the majority responded to the first question in the new category with a bid above the maximum explicit response of 110 %, a few bidding to 150 % and one to 160 %. The percentage bidding over 110 % for this question, at each site, was as follows: Site 1, 83 %; Site 2, 77 %; Site 3, 32 %; and Site 4, 92 %. The mean results of this part of the survey are presented in Table 7.12.

The other questions in this category were for:

- borrowing to attract a very large profit on the next harvest
- borrowing to increase livestock numbers profitably
- borrowing to attract a very large profit on all produce
- borrowing in order to be able to meet mortgage repayments

The first question, relating to borrowing to avert severe financial hardship, attracted the highest interest payments, followed closely by borrowing to avoid defaulting on mortgage repayments. Interestingly, the questions relating to borrowing to attract a large profit on the next harvest, and on all produce, had nearly identical responses at each site, and were 20

% to 42 % lower than for the severe financial hardship question. They were closer to the responses to the earlier WTP-PTPR and standard bidding PTPR questions, but still substantially higher.

Table 7.12 Mean WTP PTPR responses for borrowing to avert severe loss or attract large gain (% per annum ± s.e. 95% conf.)

Site number	Q 21 Severe financial hardship	Q 23 Large profit on harvest	Q 24 Increase livest'ck profit	Q 25 Large profit on all produce	Q 26 Mortgage repayment
1	117 ±2.2	95 ± 8.6	101 ±7.6	94 ±8.8	111 ±7.2
2	120 ±4.1	79 ±12.2	85 ±12.0	78 ±12.6	109±7.4
3	112 ±4.8	73 ±10.2	84±10.5	71 ±9.7	110 ±7.7
4	119 ±0.9	100 ±4.6	105 ±2.7	99 ±4.5	112 ±2.3

Livestock profitability appears to be more highly valued than general produce profitability, attracting higher WTP rates, but borrowing to avoid defaulting on the mortgage was second only to averting severe financial hardship.

Hypothesis four states that there is no significant difference between the mean WTP-PTPR when the individual is borrowing to avert a severe loss and that when borrowing to make a large gain. The relevant data from statistical testing are summarised in Table H4. The null hypothesis is rejected and the alternative hypothesis, that there is a significant difference between WTP-PTPR to avert severe financial hardship and WTP-PTPR to attract a large profit, is accepted. The WTP-PTPR to avert hardship is higher.

While this result is not in keeping with the assumption that individuals have one discount rate, it is consistent with the theory of equivalent variation (EV). Clearly, averting loss of what they have already is valued more highly by farmers than gaining something they would like to have. However, these results also make a clear statement about the socio-economic circumstances of the respondents.

Firstly, there is no reason to doubt that the initial WTP PTPR and standard-bidding PTPR responses are close to the farmers' discount rates when they have true freedom of choice.

Secondly, when asked about averting loss the farmers are prepared to pay much higher rates of interest, which reflect much more closely the revealed interest rates from their actual borrowing activities. This suggests that, in reality, they are mostly borrowing to avert some sort of loss or severe hardship (and borrowing to buy food, as so many of them are, must surely support this premise).

Thirdly, the fact that borrowing to avert hardship is more highly valued than borrowing to attract a certain profit again suggests that this is what farmers are used to doing.

Hypothesis five tests the assumption made in the first summary statement above, that the PTPR responses are close to the farmers' own discount rates, and by implication, that the WTPs expressed for borrowing to avert severe financial hardship are not. It was also suggested that this probable difference was due to the daily duress of a farmer's life.

Table H4 Summary of statistical testing for hypothesis four

Criterion	Site 1	Site 2	Site 3	Site 4
Avert severe financial hardship				
Mean WTP-PTPR (% p.a.)	117	120	112	119
s.d.	6.6	13.0	15.0	2.8
N	35	39	37	36
Attract large profit				
Mean WTP-PTPR (% p.a.)	94	78	71	99
s.d.	26.5	39.8	31.8	14.2
N	35	39	37	36
Z	5.1*	6.2*	7.4*	8.4*

*Significant at least at the 95% level of confidence

Statistical testing for hypothesis five, which states that there is no significant difference between the mean PTPR as inferred by the standard bidding question and the mean WTP PTPR when borrowing to avert severe financial hardship, is tested in Table H5.

The null hypothesis is rejected and the alternative hypothesis, that there is a statistically significant difference between the means of WTP PTPR when borrowing to avert hardship and the standard bidding PTPR in unspecified conditions, is accepted.

Table H5 Summary of statistical testing for hypothesis five

Criterion	Site 1	Site 2	Site 3	Site 4
Mean WTP-PTPR To avert hardship (%)	117	120	112	119
s.d.	6.6	13.0	15.0	2.8
N	35	39	37	36
Mean std-bid. PTPR (%)	50	44	41	46
s.d.	18	17	17	18
N	35	39	37	36
Z	20.9*	22.0*	19.0*	24.0*

*Significant at least at the 95% level of confidence

Loss and Gain, Borrowing and Lending

As a follow-up to the previous category of questions, as discussed above, three questions in the survey (Qs 22, 28 and 29) were devised to assess attitudes to loans and interest payments in cash or kind when borrowing was to avert loss or attract gain, for a range of implicit interest rates. The questions were phrased thus:

> Suppose that you stand to make a gain and a loss of equal value on your farm. You can either borrow to avert the financial loss on one commodity and in doing so forego the gain on another. Or you can borrow to make the gain on one commodity, but in doing so you incur a loss on the other. Which would you rather?

The respondent could then choose one of four responses, borrowing cash to avert loss or to attract gain, or borrowing kind (rice or corn) to avert loss or to attract gain. The value of the loan remained at 1000 pesos. The effective interest rates proposed in the three questions were 100 %, 50 % and 10 %. The difference between these questions and the ones in 4.2 is that the respondent can only break even no matter how she or he replies. The aim was to discover where the preference for borrowing for loss/gain lies and whether it changes with cash or kind loans or with interest rate.

The results (presented in table 7.13) were consistent over all sites. At 100 % interest there was a strong preference for borrowing cash to avert loss (57-69 % of respondents) followed by borrowing in kind to avert loss (19-29 % of respondents). Between 3 % and 14 % of all respondents would borrow cash to attract a gain and only at site 3 would anyone borrow in kind to attract a gain (6 %). As the interest rate dropped to 50 % the proportion of respondents borrowing cash or kind to avert loss increased slightly at sites 1 and 3 (from 86 % to 88 %, and from 81 % to 89 %), remained static at site 2 (97 %), and dropped at site 4 (from 92 % to 86 %). As the interest rate fell further from 50 % to 10 % the proportion borrowing to avert loss stayed the same at sites 1, 2 and 3 (88 %, 97 % and 89 % respectively) and rose again at site 4 (from 86 % to 89 %). The cause of this vacillation at site 4 stems from one respondent being unable to make up his or her mind. At the 50 % and 10 % interest rates no-one at any site wanted to borrow in kind to attract a gain. It does not appear that borrowing preferences (cash or kind, loss prevention or gain attraction) are strongly influenced by interest rates. The clear preference at all sites and at all interest rates is for borrowing cash to avert a loss.

The next question (Q. 27 in the survey) was designed to determine if there was a preference for short-term or long-term loans (at a fixed implicit rate of 100 % for a loan value of 1000 pesos). The periods assessed were for 1 year, 6 months, 3 months and 1 month. The preference for cash or kind was also assessed. Results are presented in table 7.14. There were eight possible responses to the question, four for cash and four for kind. At site 1, the clear preference was for borrowing over one year with respondents split almost equally between cash and kind (48 % preferring cash and 46 % preferring kind). At site 2 the preference was for cash loans, but with a wider spread over time periods, the majority (43 %) preferring three-month loans. Site 2 is the site with coconut as the main cash crop, which may be relevant since respondents at all the other sites (with rice and/or corn as the main cash crop) preferred year-long loans (94 % at site 1, 78 % at site 3 and 94 % at site 4). At sites 3 and 4 the preference was for cash over kind (70 % at site 3 and 67 % at site 4).

Table 7.13 Loss, gain and interest rates

Interest Rate % pa	Cash or kind, loss or gain	% of respondents preferring			
		Site 1	Site 2	Site 3	Site 4
100	Cash, loss	57	69	62	67
100	Cash, gain	14	3	11	8
100	Kind, loss	29	28	19	25
100	Kind, gain	0	0	6	0
50	Cash, loss	57	72	62	50
50	Cash, gain	11	3	11	14
50	Kind,loss	31	25	27	36
50	Kind, gain	0	0	0	0
10	Cash, loss	54	69	57	47
10	Cash,gain	11	3	11	11
10	Kind, loss	34	28	32	42
10	Kind, gain	0	0	0	0

Table 7.14 Preferred duration of loan

Duration of loan	Cash or kind at 199%	% of respondents preferring			
		Site 1	Site 2	Site 3	Site 4
1 year	Cash	48	8	54	61
6 months	Cash	0	13	8	0
3 months	Cash	0	43	5	0
1 month	Cash	3	15	3	6
1 year	Kind	46	18	24	33
6 months	Kind	3	3	5	0
3 months	Kind	0	0	0	0
1 month	Kind	0	0	0	0

A series of questions investigating respondents' attitudes to lending money was devised for a range of interest rates and durations to enable comparison with the questions discussed above, and to see what effect these had in the preference for borrowing or lending. There were eight questions in this series (Qs 5, 13, 19, 31, 33, 35 and 40 in the survey) and there were two possible responses to each question: lend or borrow. The questions each gave a specified interest rate for a given period and a stated amount of money. The interest rates were 10 % and 50 % per annum, the durations were one year or three months, and the cash amounts were 1000 pesos or 200 pesos. The results to these questions are shown in Table 7.15.

Table 7.15 Lending and borrowing with variation in interest, duration and value of loan

Int. Rate % p.a.	Am't pesos	Duration	Lend/ Borrow	% of respondents preferring			
				Site 1	Site 2	Site 3	Site 4
10	200	1 yr	Lend	9	18	11	8
			Borrow	91	82	89	92
10	200	3 mth	Lend	17	15	22	3
			Borrow	83	85	78	97
10	1000	1 yr	Lend	3	5	14	3
			Borrow	97	95	86	97
10	1000	3 mth	Lend	3	3	16	3
			Borrow	97	97	84	97
50	200	1 yr	Lend	63	72	76	64
			Borrow	37	28	24	36
50	200	3 mth	Lend	63	82	70	50
			Borrow	37	8	30	50
50	1000	1 yr	Lend	54	64	65	64
			Borrow	46	36	35	36
50	1000	3 mth	Lend	48	67	62	58
			Borrow	52	33	38	42

As might have been expected, a very large majority (82 % to 97 %) would prefer to borrow at 10 %; for them this is a very low rate of interest, and most would never have the option to take advantage of it. Most would also (understandably) rather borrow 1000 pesos than 200 pesos at this rate. The duration of the loan (one year or three months) did not have a strong influence on the result. However, when the interest rate was 50 % in most cases a smaller majority (50 % to 76 %) preferred to lend the money, rather than borrow it, although more preferred to lend 200 pesos than 1000 pesos, with more preferring the shorter period of 3 months (there was some variation between sites).

Summary

This study has covered a wide range of issues relating to site characteristics, income, equality, tenure, borrowing, and interest and discount rates of Philippines upland farmers. The study sites are not homogenous in any socio-economic sense, and there are numerous financial and behavioural differences among respondents within and between sites.

The rate of adoption of soil conservation practices varies greatly between sites as does the rate of borrowing among soil conservation practitioners, with practitioners having a slightly lower borrowing rate, and when they borrow, borrowing less money. It is difficult to establish a link between borrowing and soil conservation practices, although at is just as difficult to conclude that no link exists. However, when we investigate the observed interest rates paid by borrowers, and the high perceived discount rates of respondents, it is not surprising that people with such a clear need to consume in the present (or not to defer consumption) do not have a high rate of soil conservation adoption, given that they would receive no benefit from such practices for several years.

The results of hypothesis testing so far lead to the conclusions that:

- There is a significant difference between mean interest paid on debt and farmers' mean private time preference rates. One should not be used as a shadow for the other at the study sites.
- At sites 3 and 4 there is a significant difference between mean interest rates charged on cash and in-kind loans (in-kind loans are dearer).
- There is a significant difference between mean WTP-PTPR to avert financial hardship and mean WTP-PTPR to attract a large profit (the former is higher).

- There is a significant difference between farmers' perceived mean standard bidding private time preference rates and their mean WTP private time preference rates to avert financial hardship.
- There is no significant difference between mean WTP-PTPR and standard bidding PTPR results when the data are aggregated and testing using a regression analysis; however there is a varying degree of difference using site specific z tests.

Interesting general observations arise from analysis of the surveys discussed here, and these results may provide a challenge to some assumptions about the use of conventional economic theory in policy development. A brief summary of results is presented below:

- Socio-economic conditions at the study sites are by no means homogeneous. When data were analysed on a site-specific basis, results sometimes varied (such as the link between mean WTP PTPR and mean standard-bidding PTPR, and the link between soil conservation and mean perceived discount rate). Policy makers must always be aware the conditions may vary from site to site, and policy probably should be developed on a case-by-case basis.
- The discount rate of farmers, as reflected in their willingness to pay, varies according to the condition of the loan.
- A large proportion of respondents borrow money to buy food.
- Observed interest rates paid on borrowings at all sites are extremely high, and bear little relationship to the WTP-PTPR and standard bidding PTPR perceived discount rates elicited in the survey, when the purpose of the loan is not made explicit.
- Duration of loans appears to have a greater effect on WTP than on value of loans, but varies with both. WTP also varies with cash or kind.
- When WTP is elicited for loans to avert severe financial hardship the perceived discount rate is much closer to the observed interest rate.
- WTP-PTPR for loans to attract profit is higher than, but closer to the WTP-PTPR and satndard bidding PTPR for unspecified loans.
- Respondents show a strong preference for borrowing to avert a loss rather than attract a gain (when these will cancel each other and the net result is equal).
- Most respondents prefer to borrow cash rather than kind, but as interest rates fall more will take in-kind loans (although most still prefer cash). However, most express a greater WTP for in-kind loans
- As interest rates fall there is a slight increase in borrowing for gain, though most still prefer to avert loss.

- There is a very strong preference for long-term cash loans over short-term cash loans except at site 2 where kind is preferred at 100% over a year. None at any site prefers kind over cash for 1 and 3 month durations.
- Respondents prefer to borrow at low interest rates and lend at higher interest rates. They also prefer to borrow in the long term and lend in the short term.
- The greater the interest rate the greater the preference for lending over borrowing.
- Respondents prefer to lend small amounts and borrow large amounts.

Notes

1.   The observed mean interest rate paid on all borrowings at each site is the mean of all categories of loan, whether cash or kind, for all purposes and all durations, and for small and large amounts. Weighted mean interest rates are the means of all loans weighted according to the value of the loan.
2.   This is at odds with what might reasonably have been expected, since theoretically, smaller loans attract higher interest rates because they have higher transaction costs.
3.   But according to economic theory should reflect the individual time preference rates of the borrowers, in a functioning capital market.
4.   However, for reasons that are not clear, when the sites are examined individually, there is a difference between WTP PTPR and the standard-bidding PTPR at sites 1 and 2.
5.   Borrowing activity may well be a function of need and creditworthiness, except at site 2 where the reason for a different borrowing pattern is unclear. We can only speculate about the cause, but it is possibly related to the insurgency problem at the site. The insurgency problem had led to the formation of a vigilante group which required membership of able-bodied men, and raised a levy on the villagers for the purchase of food, arms and equipment.

Chapter 8

Implications of the Results
for Theory and Policy

Introduction

This book has covered a wide range of topics between which there is a network of links. It is not possible to attempt to explain the socio-economic implications of the upland farmers' decision making processes without considering the physical and cultural influences on their lives. Martinez-Alier (1990, p. 160), who has written a far-reaching book on ecological economics, has commented on the interdisciplinary breadth required to investigate the current attitudes and approaches to understanding patterns of resource allocation. He states, when commenting on discounting and time preference, and on the inter-generational allocation of exhaustible resources (such as soil and old growth forests):

> One possible line of work would have led to writing an essay in moral philosophy or in moral exhortation, or perhaps, more fruitfully, would have led to an investigation of the social distribution of different moral values on intra- and inter-generational economic equality (do they correlate with social class, income level, age, rationality, religion, gender?). In fact, the economist, in order to explain the historico-geographical allocation of exhaustible resources (and industrial waste), which is a major part of economic history over the past two hundred years, must become, certainly not a moralist, but a sociologist and a historian of the ethics of time preference.
>
> Economists should also become students of the history of science and technology, since economic agents will take their beliefs about technical change from this history (from where else?).

In addition to the fact that much analytical and policy economics lacks an historical perspective, the implications of distributional effects are also frequently overlooked in policy development. For this reason an assessment of income distribution at the study sites (see Chapter 6), which builds on the discussion of social and economic history (see Chapter 2) and on the socio-economic profiles of the study sites and the Philippines (chapters 3 and 4), has been included. Given the results presented in chapters 6 and 7 these perspectives should not be overlooked, especially since the assumptions in standard economic theory do not adequately cover the divergences that we witness in some of the survey responses. For example, when respondents were asked to express WTP for cash or in-kind loans, the higher interest rates were proposed for in-kind loans from which, according to theory, we should infer a preference for in-kind over cash loans. However, when respondents were asked directly if they would *rather* take cash or in-kind loans the unambiguous preference was for cash loans. In addition, the extremely high interest rates which respondents are paying for food and medicines is clearly having some undesirable effect on resource allocation such that projects with significant long-term social and economic benefits (e.g. soil conservation) are being foregone.

Other researchers have also found inconsistencies between the assumptions about time preference rates in economic theory, and the behaviour of poorer people in less developed nations. For example Moseley (2001, p. 321), when assessing the relationship between poverty, time preference and the environment in Africa, comments: '... these theoretical justifications are culturally laden assumptions that may make sense in a Western context, but are more problematic in many rural African settings, particularly in the poorest and most food insecure contexts.' The lack of recognition of the relevant socio-cultural and historical context of decision makers, in developed and less developed countries alike, may be leading to misguided social and environmental policy, and further exacerbate the unsustainability of food production practices. Moseley (*ibid*, p. 325) concludes:

> Further study by economists in the behaviour of food insecure households in rural Africa may help the academy and policymakers better understand rates of time preference of the poor. While this article has undoubtedly presented more questions than answers, it has hopefully highlighted problems with current theory and policy and given cause for reassessment of both.

There are obvious distributional considerations which should be taken into account in policy development. Argy (1993, p. 98), when speaking of the poor image of economists states:

Another factor which has contributed to the erosion in our standing and credibility has been our failure, in general, to allow systematically, for distributional effects in our overall evaluation of policy.

Economists often seem to argue as if any structural reforms which improve economic efficiency must of necessity improve the economic welfare or total well being of the nation.

The case is unambiguous where the reforms are of the win-win (Pareto improving) variety ...

But the majority of reform strategies involve (at least in the absence of compensatory action) a great number of losers as well as winners.

Argy (*ibid*, p. 103) concluded his address by asking for a more multidisciplinary approach to economics:

Economists will also need to draw increasingly on ideas and perspectives from other social sciences if they are to make their voices heard in the increasingly competitive market for policy advice.

The results of the socio-economic surveys presented in this book, and particularly those relating to borrowing activities, will be examined in the light of of what is known about the historical, social and cultural perspectives of the respondents. For example, religion, especially the Roman Catholicism introduced by the Spaniards, is very important to most of the villagers. The influence of the Church was very much stronger than the influence of the State or the Crown for several centuries after colonisation, and it is probable that some of the borrowing for food is to enable purchase of expensive items (such as roasting pigs) for Saints' days and other religious festivals which have great significance to many of the respondents.

Another example: anecdotal evidence suggests that the very high interest rates charged at the study sites might not be explained to any great extent by lenders' risk, because default rates on loans from usurers are very low. However, it is also possible that default rates are low at the study sites (around 5 %) because of the intensive follow-up system used for debt collection (money lenders often collect the debts in person as the harvest is sold). Because of the system of forfeit (possibly introduced by Chinese traders), the risk to borrowers is often much greater than the risk to lenders, since debtors unable to pay the required amount on time (often a lump sum of capital and interest) stand to forfeit their land, if they have any, and significant other property if they do not. Local people in Leyte said that often recruitment to the NPA had less to do with political ideology than with indebtedness.[1] Those who had forfeited their land sometimes had nowhere to go and no means of earning a livelihood and so 'went to the hills' (the local euphemism for joining the NPA). The author was also told that in one part of Leyte a Catholic priest was the biggest local landowner, landtaker and usurer. His motivation, we were assured, was to consolidate a decent inheritance for his children whose mother was his housekeeper.

While most of the above was provided as anecdotal evidence, such phenomena are widely recognised and sometimes catalogued in the Philippines (see Floro and Yotopoulos, 1991). It is credible in the light of the history and culture of the place, and gives some plausibility to unexpected and otherwise seemingly inexplicable observations which may seem at odds with economic theory. However, if the theory itself holds well enough, the assumption of a free capital market in the local study site economies does not.

In other parts of the Philippines and in other less-developed countries in general, evidence on default rates is mixed. Default rates on loans from co-operatives and overseas aid development projects can be as high as 10 %, often up to 50 % and sometimes even higher. For such loans default penalties are not dire and interest rates (much lower than usurers rates) may have a built-in lenders' risk component. The European Economic Community has factored a 10 % default rate into interest rates applied to its overseas aid projects.

Assessment of Results in the Overall Scheme

As discussed earlier, there are large socio-economic differences between residents within each study site, and when the sites are compared with each other. The income distributions are highly skewed, as can be seen when the mean and median incomes at each site are compared, and when the income inequality ratios and Gini coefficients are examined.

The poorest site, at the income-mean and total income level, has the most equal income distribution, and the richest site has the least equal income distribution. Patterns of settlement and land tenure vary from site to site. Pomponan (Site 2) is the oldest site and has been settled for many centuries, which has led to it having the most degraded soils. Although land ownership is relatively high at this site, income is low.

Tabing (Site 4) had a high rate of amortising owners due to being party to a land reform program where hacienda tenants were able to purchase the land they tilled. Yet Tabing respondents have the lowest mean and median incomes. Historically this is unsurprising for, as Uy Eviota (1992, p. 100) states:

Smallholding households became precarious units with the increased commercialisation of the village economy; competition forced many out of the market. Some lost their land as they failed to meet mortgages taken out to finance capital outlay on the farm. Others were dispossessed of their land by agribusiness interests. By 1975 there were three-and-a-half million landless farm workers, which was still 47 % of all agricultural workers and about 25 % of the total labour force...

An agrarian reform program implemented at this time also adversely affected landless labour and small farms. Typically of land reform programs, this one was limited in scope and failed in the main aim of reform – the redistribution of wealth. The high valuations of land in the transfer of ownership to tenants and the increased cash outlay needed for the high-yielding crop varieties resulted in a growing debt for many farms.

If we re-examine the study site profiles (see Table 7.2) we can see that the rate of adoption of soil conservation varies from site to site, and that the tenancy status of farmers coupled with the relative wealth of the site (as indicated by median income levels) appears to influence this rate. Site 2, although poor, has the highest rate of soil conservation adoption and the highest rate of outright land ownership. In contrast site 3, with the second-highest median income (and the highest mean income), has the lowest rate of soil conservation adoption (only 10 %), and the highest tenancy rate.

Yet site 1 has a high tenancy rate, the highest median income by far and the second-highest soil conservation adoption rate, even though it has the lowest level of erosion. At site 1 the mean and median interest rates are the lowest of all sites, while the WTP and standard-bidding PTPRs as evinced by the 1993 survey, are the highest. And when willingness to pay for credit is assessed on the basis of specified financial loss and gain questions, site 1 is the only site where the WTP is roughly equivalent to the median interest rate actually paid. At the other sites WTP is always a lot lower than the observed site interest rate.

This gives some credibility to the idea that the adoption of soil conservation practices is influenced by the relationship between the farmers' discount rate and the observed interest rate on borrowings, in addition to the influence of income and land tenure, and other factors such as age and education level (all of which influence the discount rate).

Sustainable Resource Use

Much has been written about sustainable resource use and sustainable economics. It seems that land-use in the Philippines (and, if we consider the issue country by country, worldwide) is not sustainable. We are already witnessing the consequences of deforestation and cultivation of unstable slopes, and the destructive influences of cyclones, floods, landslides and tidal waves – which have a very high social and financial cost (see Chapter 3). Standard economic theory casts the decision maker in the role of 'rational economic man' and argues that neither individuals nor firms, nor governments, will make decisions that cost them rather than benefit them. Unfortunately these identified groups operate often from different perspectives but never in isolation from one another, and what is rational at one level may be irrational at another. We have seen that this study's respondents do not always adopt soil conservation technologies which will have long-term benefits for them and their descendants. The results of the surveys indicate that from their own perspectives these farmers are acting rationally. In the light of their social, physical and cultural history they are making sensible choices in so far as they have any choice at all.

Jacobs (1993, pp. 25-26) compares government policy making with strategic development in industry. He states:

> In industry it is now widely accepted that you should start off with a mission statement and then set yourself strategic targets for a five or 10 year period (where you want to be in terms of sales size, employment, different markets you want to be in); and then you should look at how you are going to get there, your strategies: what you need to do in terms of human resources, investment, marketing, and so on.
>
> It seems to me that the environmental policy-making process, but much more importantly the economic policy making process, should take a rather similar approach. We should say, as we would in industry, this is where we want to be in 10 years' time. These are our objectives measured in terms of chosen performance indicators – this is our employment objective, this is our income onjective, these are our environmental objectives. Then, given these objectives, how do we meet them? What do we need to do, and how can we implement it?

If any government is serious about sustainable economic development and sustainable resource use, remembering that interpretation of these concepts is a very personal thing (McManus, 1996), it will have to devise and implement clear policy on the matter. And governments are going to have to reconsider the idea that the market will allocate environmental resources and natural resources efficiently, and that all decision makers have the unencumbered freedom to choose. Our respondents are aware of the need to conserve their soil; it is, after all, the source of their livelihood. The upland farmers are keen to consider future generations, they think about their children and their grandchildren; yet their discount rates appear to be very high. Most of them cannot defer consumption because, if we look again at our purpose-of-borrowing responses, this may mean deferring eating, planting, and even buying medicine and burying the dead. Individual farmers also have almost no control over what happens higher in their catchments. The upland degradation of the Philippines is mostly as a result of logging and cultivation of the steep unstable slopes, over which the government has some control.

Governments, and even aid organisations, should develop policies in the light of facts. Some people are probably going to need economic help for the good of society. The form this policy development should take will be discussed later in this chapter.

Policy is often theory-driven, and may be postulated in ignorance of what is actually happening. This chapter shortly will explore the consequences of this approach, and will compare and contrast the theory with the reality in view of what theory says of the links between interest rates, discount rates and decision making. Meanwhile, theorists and practitioners are pursuing some of the issues that may drive sustainable resource use policy (see Sharma, McGregor and Blyth, 1991; Sakar and McKillop, 1991; Van den Bergh and Nijkamp, 1991; Resource Assessment Commission, 1992; Ekins and Max-Neef, 1992; Van Pelt, 1993; Young, 1993; Bowers, 1997; Buttel, 1998; Pearce, 1999 and Ullston and Rapport, 2001).

The causes and effects of land degradation, especially in developing countries, involve a complex network of links which makes it easy for policy makers to shift responsibility from themselves, and makes it difficult to target any individual issue. Barbier (1991, pp. 76-77) states:

> For environmental degradation to be an economic problem, however, requires it to have a cost in terms of human welfare, either directly or indirectly, currently or over time.
>
> Perhaps the most difficult feature of environmental degradation is that its costs often occur externally to any market system and involve complex processes of ecological-economic interaction. Both the economic causes and the effects of environmental degradation are difficult to discern and analyse in developing countries.

As the severe consequences of land degradation are felt more widely around the world, governments of developed and developing countries, and international aid agencies, are seeking answers that will treat the cause, rather than the effect of the problem. Barbier (*ibid*, pp. 83-84) continues:

> Appropriate incentive structures must be set against the context of management regimes – access and rights to land and other resources – and the complexity of poverty-environmental degradation linkages in developing countries

and further ...

> As the environment deteriorates, poor rural households are constrained to shorter time horizons and restricted choices. The increased risk and uncertainty over livelihood security will tend to make poor households reluctant to change their farming practices, grazing patterns and other economic activities, and in the short run, these changes are perceived as adding to the risk burden.

This is very much the crux of the research presented in this book. It is unlikely that ignorance or even resistance to change prevents the Leyte upland farmers from adopting soil conservation practices with long-term benefits. As mentioned earlier, they already face risks from natural disasters, political instability, accidents, disease and malnutrition; they are reluctant to add another risk category. In addition, as shown in Chapter 7, their high personal time preference rates are, in most circumstances, considerably exceeded by the interest rates that they **must** pay on debt. It is almost inconceivable that the majority of farmer respondents could implement soil conservation without borrowing for the purpose.

As shown earlier, anecdotal evidence tells us that at these sites, contrary to popular belief, borrowers' risk exceeds lenders' risk. If the forfeit is one's livelihood then one would be irrational to stake it on a loan with mainly long-term benefits for the land that would be forfeit if the loan were in default.

Barbier (*ibid*, p. 100, p. 105) moves towards incentive proposals as a solution to such problems. The research findings in this book support his view that:

> More complex incentive effects arise from the relationships between erosion and the availability of labour, off-farm employment, population pressure, tenure and access to frontier land, the development of post-harvesting capacity and other complementary infrastructure, and the availability of affordable interest rates.

He concludes ... 'The ultimate objective must be to develop appropriate incentives for private decision makers.' There is some consensus among resource economists. The range of options for policy development is running low. The 'free market' solution appears to be failing, limited in part by the farmers' lack of freedom to make choices. The tragedy of the commons is fast becoming the tragedy of the alienated lands.

Turner (1991, p. 175) under the heading of 'Key Intervention Points for Sustainable Development' tops his list with:

> (a) The household/community level is critically important and offers considerable scope for intervention channelled through non-governmental organisations (NGOs). The provision of credit for small farmers is a high priority here.

This study has focussed on the household and community level in an attempt to assess the process by which individual farmers make decisions about land-use practice. When the focus is on the issue of rural credit, and when it is apparent how large the divergence between PTPR and interest on debt really is, the importance of the provision of affordable credit becomes clear.

The Divergence Between Theory and Reality

The hypothesis that the relationship between interest and discount rates holds in the manner proposed by standard economic theory has been examined. That is, that the market interest rate equates with the private time preference rate. This theory is outlined by Fisher (1965, p. 104) as follows:

> ... the rates of time preference for different individuals will, by the process of borrowing and lending, become perfectly reconciled to the market interest rate and to each other, for if, for any particular individual, the rate of preference differs from the market rate, he will, if he can, adjust the time shape of his income stream so as to bring his marginal preference rate into harmony with the interest rate.

Layard (1980, p. 32) states that ... 'In a mixed economy the most obvious indicator of time preference is the market rate of interest.' Layard (*ibid*, p. 35) comments that the uncertainty of future capital markets leads to a range of interest rates which reflect the risk of lending and argues that:

> This is why the most commonly used estimate of private (and social) time preference is the rate on long-term government bonds (reduced to allow

for expected inflation and income tax payments). These bonds are generally considered to be risk free.

Thus Layard not only suggests that in standard theory the interest rate is a reflection of the individual's discount rate (PTPR) but that this is, in turn, a reflection of the social time preference rate (STPR). However, the interest rates available to Leyte upland farmers are likely to be offered in a range of imperfect markets rather than in one perfect market.

A range of interest rates at the study sites have been observed, but we could assume from the above that they reflect the market rates at these sites. We could assume also that they at least approximate PTPR, but analysis of the results has shown that they do not.

This study has also demonstrated that individuals' private time preference rates vary according to the conditions under which credit is offered. As reported in Chapter 7 there was a great difference between WTP for a 1000 peso loan when no particular purpose was expressed, and when the loan was specifically to avert a severe loss or to attract a large gain. These differences were in the range of 30% per annum to 78% per annum. The highest expressed willingness to pay for a loan was to avert a severe financial loss. Yet these WTP results were still less than the observed median interest rates, although the gap at site 1 was only 3%, while at site 4 it was 110% per annum. In addition, individuals expressed different WTP-PTPRs when the amount, duration and physical nature (cash or in-kind) of the loan was varied, although this variation was much smaller (up to 15% per annum) than that shown with specifying the purpose of the loan for an extremely good outcome or to avert an extremely bad outcome.

The hypothesis that the farmers' private time preference rates approximate the interest rates they pay on loans has been rejected. Part of the reason for this non-approximation of interest rates with time preference rates is that the mean PTPR applies to the whole sample at each site, while mean interest rates only apply to borrowers, who constitute about 50 % of each sample. However, this is only a minor contributor to the difference between interest rates and PTPRs. Results also indicate that the PTPR of farmers is not static at any given time, but varies according to the purpose of the loan, the duration, amount and the cash or in-kind nature of the loan. These variations do not appear to be dependent on the income of the farmers, though they do seem to be influenced by the farmers' expectations of loss or gain. These divergences are probably strongly influenced by the range of risks that farmers face in their day-to-day decision making, and the dire consequences of misjudging the risks. As Layard (*ibid*, p. 54) points out … 'Much of human behaviour towards risk can be explained by the hypothesis that people maximise their expected utility.'

Clearly, if the harvest should fail, the land be forfeited or the family go hungry, there will be a drop in utility. Also, the divergences probably are strongly influenced by the farmers' lack of choice – standard theory assumes freedom of choice in decision making within an open capital market, but farmers are constrained by their lack of choice with respect to whether or not they borrow, and what interest rates they pay on borrowings. In addition to the above observations, we noted some other unexpected results from the surveys. When assessing WTP and PTPR by survey we tend to make assumptions from the magnitude and direction of results. For example we would assume that a higher WTP for a certain good would reflect a greater preference for that good over another for which WTP is lower. However, when the WTP responses for cash or in-kind questions were tested against a qualitative 'would you rather?' question the results were surprising. Generally respondents expressed a higher WTP for in-kind loans, yet when asked 'would you rather cash or in-kind loans?' most expressed a preference for cash loans. We must be careful in how we interpret WTP survey responses, since there are many (often unknown) influences on replies, and clearly a higher WTP does not always signify a greater preference.

With respect to the range and magnitude of interest rates paid by farmers, it is possible that there is a multiplicity of unconnected credit markets operating. It would be impossible to merge these to make one market, so policy makers should recognise the different markets and accommodate them on a case-by-case basis.

Some of the results are puzzling. Usually it is accepted that individuals with high incomes have a lower time preference rate than those with low incomes. However, at site 1 respondents have the highest median income of the four sites yet they express the highest PTPR for most categories of survey question. Respondents at site 4 have much the lowest incomes yet they express the lowest PTPR for most questions.

All of the factors described above need to be taken into account by policy makers if there is a serious wish to influence farmers' activities, particularly with respect to sustainable resource use.

Policy Implications

The results of this research clearly carry a message for policy makers. The range of possible influences on the rate of soil conservation adoption among the upland farmers in Leyte has been assessed. This assessment has concentrated particularly on income and credit effects. Parrilla (1992), using the same samples, examined the broad range of socio-economic indicators and used econometric analysis to determine the most likely influences on the adoption of soil conservation technologies. She found that liquidity, age, education, the perception of erosion, land tenure and 'discount rate' all had a significant effect on adoption practices. 'Discount rate' is included in quotation marks because Parrilla used the interest rate as a surrogate measure for the discount rate. She states (p. 167):

> Discount rate is expected to influence strongly the adoption of soil conservation practices in the upland. However such data is not easily collected and measured in personal surveys. Therefore the imputed interest in this study is used as a surrogate measure for the discount rate. The higher the discount rate, the less likely is the farmer to adopt the technology.

As we have discussed above, the interest rate, in this study, does not prove to be a reliable surrogate for the discount rate of the farmers. The PTPR, when estimated from using the standard approach to discerning PTPR, is actually very close to the WTP for interest on the same amounts of money over the same duration (see Table 8.1 for a summary of mean and median interest, discount, debt and income information). Yet these figures are well below the observed interest rates, which are more closely shadowed by significant gain or loss WTPs. In addition, when we examine the discount rates estimated from the perceived discount rate survey we can see that there is not a simple, direct relationship between discount rate and soil conservation adoption. Nor is there a simple, direct relationship between interest paid on loans and soil conservation adoption. It is tempting to try and distil explanations for human behaviour into a single factor, but this represents an oversimplification.

In Table 8.1 (below) the more apparently important determinants of adoption of soil conservation technologies (and other forms of sustainable agriculture) have been summarised. Parrilla (*ibid*) conducted a broadly based and sophisticated regression analysis (as discussed earlier) and found that a range of factors weakly influenced adoption practices.

However several important observations of the data can be made. These are as follows:

- Two of the sites have tenancy as the predominant tenure status. One of these sites has a much higher income than the other three sites.
- The site with the highest rate of soil conservation adoption has the highest proportion of outright owners at the site. It is the second-poorest site.

- The site with the (far) lowest rate of soil conservation adoption has the highest tenancy rating, the highest level of income inequality, the lowest proportion of borrowers in the sample and the lowest mean PTPR.
- The two sites with the higher rates of soil conservation have access to formal credit sources (the other sites do not), and have higher median loan values, a higher number of borrowers in the sample and higher median-loan-value/median-income ratios than the other sites.

Careful examination of Table 8.1 leads to the conclusion that tenure status, income level, access to fair credit and the discount rate of the individual concerned are indicated as the main determinants of soil conservation adoption. Recall that even when formal credit is attracting a high interest rate, willingness to borrow will be higher because penalties for default are not nearly so severe as with usurers. The gap between individual discount rates and interest rates is also likely to have an effect on borrowing.This in turn will affect adoption of conservation technologies. Respondents at site 4 have a large proportion of amortising owners, very low incomes, low loan values and very high interest rates on credit, yet they have a higher rate of adoption than respondents at site 3, who have a higher median income and lower credit charges but 79 % of respondents who are tenants. Conversely, site 1 has 75 % tenants but a relatively high income, the lowest credit charges, the second-highest rate of soil conservation adoption and the smallest gap between the interest and discount rate.

It seems as though the poorest owners, even with inadequate access to fair credit, will attempt to use conservation measures, while well-off tenants with access to fair credit also will attempt to employ conservation measures. The 90 % failure-to-adopt rate is the lot of poor tenants with no access to fair credit. Results also suggest that fair credit access leads to higher borrowing rates with higher loan values, and this seems to be linked to greater adoption of soil conservation.

Thus if policy makers are to maximise social benefits, they need to develop policies which focus on several factors at once. These are land tenure, income, and farmers' time preference rates and access to fair credit. These factors form complex interrelationships and are probably relevant for all nations, developed or otherwise where land degradation is a problem.

Table 8.1 Summary of important socio-economic indicators

Indicator	Site 1	Site 2	Site 3	Site 4
Mean obs. int. (%)	110	176	178	409
Median obs. Int. (%)	118	116	143	436
Mean Std-bid. PTPR (%)	52	43	41	44
Mean WTP PTPR (%)	64	50	45	40
Mean WTP PTPR % purpose spec.	104	94	90	107
Mean income (pesos)	18,765	11,167	30,291	6902
Median income (pesos)	15,021	9396	8349	6794
Mean brwr income (pesos)	16,972	11,360	9740	6096
Median brwr income (pesos)	14,718	8550	8914	7201
Mean loan value (pesos)	1094	1209	1031	241
Median loan value (pesos)	1600	1135	810	271
Mean loan/income %	6	11	11	4
Med. Loan/income %	11	13	9	4
Predom. Tenure %	Tenant (75)	Owner (56)	Tenant (79)	Amort-owner (71)
Soil cons. adopt. %	43	50	12	42
Soil cons adopt, brwrs %	56	67	40	61
Brwrs in sample %	67	67	47	62
Formal credit	Yes	Yes	No	No
Gap between mean PTPR and mean interest paid %	58	133	137	365

The site where the median interest rate approximates the median willingness to pay for a specified purpose has a median loan value much higher than the mean loan value, even though the median income of borrowers is lower than the mean income of borrowers. At the other sites, where the gap between median interest rates and median WTP for a loan for a specified purpose is very large (90-229 %), the median loan value is close to, or in some cases lower than, the mean loan value. At two sites with the rates of soil conservation adoption close to 50 % the median loan-to-income ratio is higher than the mean loan-to-income ratio (at site 1, the site with the best credit access, it is nearly double). At site 3, the site with the lowest rate of soil conservation adoption, the median loan-to-income ratio is lower than the mean loan to income ratio. This suggests that where relative income is less of a constraint, borrowing rates and adoption rates are higher.

In the light of the results on the differences in perceived discount rates when farmers are borrowing to avert a loss, compared to when they are borrowing to attract a gain, it would seem that they are far more sensitive to large losses than expected, relative to how they view large gains. This dichotomy probably results from their perception of the risks involved when carrying losses. They are prepared to pay a high price to avert such losses, since the inability to meet outstanding debts may involve the forfeit of their livelihood. Gains, while very attractive, are likely to be viewed as a windfall, and must take second place to averting loss. This result is in keeping with the observation that while willingness to pay for something we do not have is, in theory, equivalent to the amount we wish to be compensated for the loss of something we already possess, in reality the amount required for compensation is usually much higher.

Thus it is highly likely that a site with high land ownership levels, increased income and fair credit would demonstrate higher levels of soil conservation adoption. However, at this stage we do not know what level of return on their investment most respondents would require before adopting soil conservation. From a policy perspective this would influence the level of subsidy to be offered. Indications are that the required return on investment may be as high as 100% (Jayasuriya, 1994). Clearly, non-sustainable land-use practices are going to have long-term social costs and it will be in the interests of governments in developed and developing countries to encourage long-term projects which encourage sustainable resource allocation.

Only well planned agrarian reform programs will address the issue of land tenure, though enlightened landlords will certainly improve the lot of tenant farmers. Unfortunately we lack information on the relationship between landlords and their tenants at these study sites, and so cannot comment on their roles in the apparent differences in the rates of soil conservation adoption among tenants (as at sites 1 and 3). Since the respondents in the project survey samples are aware of the benefits of adopting soil conservation technologies, education programs will be of limited benefit.

At sites where the predominant land tenure is that of tenant or share farmer, the landlords' attitudes to soil conservation are likely to influence the rates of adoption. Because we do not know the landlords' attitudes to their land we cannot assess the type of policy required to influence landlords. It is possible that the relatively high rate rate of soil conservation at site 1 compared with site 3 (both have a large number of tenant farmers) may in part be influenced by landlords' attitudes and their assistance (or lack of assistance).

If landlords view their tenanted property much as they would view a capital asset that is written off over a number of years (as some businesses might view a tractor, factory plant or office equipment) then they are unlikely to be prepared to invest in long-term conservation technologies. It is also possible that the landlords might view the land as a political (rather than productive) asset (Anderson, 1994). A political asset would give landlords political or feudal power and may influence holding of office. If this were so, there would be no strong incentive to assist with influencing the long-term productivity of their land, since ownership and political influence are unlikely to be altered by investment in soil conservation technology.

The problem of low income among farmers could be addressed by various measures, depending on its cause. Probably the most easy issue to target is that of fair credit. There has been little research on the relationship between sustainable land management and interest and discount rates, and this makes policy determination more difficult because it is hard to assess whether a consistent pattern is emerging, especially in less-developed countries.

Pender and Walker (1990) conducted hypothetical and game-theory surveys in India in an attempt to determine time preference rates of villagers. Their findings were not entirely consistent with those presented here. For example (p. 28):

> Given a mean per-capita income of about $200, high rates of discount were anticipated. This expectation was fulfilled as mean discount rates ranged from 0.3 to 0.60 depending on the experimental game and the hypothetical question. The estimated discount rates were significantly greater than the average interest rates on debt outstanding.

This result is at odds with the findings in this book, where the perceived discount rates were usually much lower than the weighted mean and median interest on debt outstanding. However the following statement (*ibid*, p. 29) warrants agreement:

> marked time and magnitude effects, documented in the within-subject analysis of variance, are consistent with much of the experimental literature on time preference. These effects suggest that the violation of the axioms of the discounted utility model is a common occurrence.

In a study of decision-making among farmers in Victoria, Australia - a study with the goal of providing information for policy makers – O'Brien (1987, p. 2) concluded that ... 'presentation of the 'right' and 'relevant' information is necessary but may not be sufficient to change a farmer's behaviour', and that the ... 'economic objective is aimed at trying to balance the books, of trying to earn sufficient income to pay living expenses, operating costs and the oppressive interest bill'. Thus even in relatively affluent Australia there are strong parallels with the influences on farmer decision making in the Philippines.

Results presented here indicate that the perceived discount rates of individual farmers, as determined by their private time preference rates, have an effect on soil conservation adoption. In Table 8.2 the discount rates of adopting and non-adopting farmers are compared as are observed interest rates paid by them on outstanding debt. These data are used to test hypotheses six and seven, previously outlined in Chapter 7.

Hypothesis six states that there is no significant difference between the mean discount rates of respondents who adopt soil conservation practices and the mean discount rates of those who do not.

Hypothesis seven states that there is no significant difference between the mean interest rates paid by those who adopt soil conservation and the rates paid by those who do not.

At all sites but site 1 the farmers adopting soil conservation practices have a lower perceived discount rate than non-adopting farmers, when the mean standard-bidding PTPR for 1000 pesos over one year is compared. The median amount for standard-bidding PTPR at all sites is lower for adopters. It is not clear why the mean for site 1 is higher for adopters. However, for WTP-PTPR criteria for 1000 pesos for one year both the mean and median values for adopters at all sites are lower than for non-adopters. This is an interesting result; however it is consistent with theory, since those with a lower discount rate value the future more highly than those with a high discount rate, and are therefore more prepared to invest in a technology with long-term benefits.

In the case of observed interest rates (the interest they are paying), adopters pay lower interest at the mean and median at all sites but site 2, where for some reason non-adopters pay slightly less interest. Given that borrowers do not represent the whole sample, this (interest paid) is not a very reliable indicator of the sample willingness to pay, nor is it a good indicator of the individuals' perceived discount rates for reasons already discussed. In addition, as seen from the standard deviations, there is a large amount of variation in the rates paid, and where there is a small number of borrowers in the relevant sample the results may not be reliable.

Table 8.2 PTPR, WTP, interest rates, for adopting and non-adopting farmers (A=adopters; NA=non-adopters)

Criterion		Site 1	Site 2	Site 3	Site 4
Std-bidding PTPR 1000 pesos/yr					
A	Number	16	20	5	13
	Mean PTPR (%)	61	44	35	38
	s.d.	26	17	14	20
	Median PTPR (%)	50	25	27	38
NA	Number	19	19	35	22
	Mean PTPR (%)	47	45	42	50
	s.d.	20	16	19	15
	Median PTPR (%)	50	50	50	50
WTP PTPR 1000 pesos/yr					
A	Number	16	20	5	13
	Mean PTPR (%)	55	49	20	33
	s.d.	27	17	10	20
NA	Number	19	19	35	22
	Mean PTPR (%)	64	51	52	47
	s.d.	21	14	37	19
	Median PTPR (%)	60	60	30	60

Table 8.2 (Continued)

Criterion		Site 1	Site 2	Site 3	Site 4
A	Number	11	15	2	7
	Mean int. rate (%)	116	205	151	384
	s.d.	60	67	13	168
	Med. int. rate (%)	87	151	151	410
	% borrowing	65	71	40	50
NA	Number	15	11	17	14
	Mean int. rate (%)	135	179	183	402
	s.d.	147	125	91	267
	Median int. rate (%)	119	150	171	463
	% borrowing	68	58	48	58

When conducting statistical tests on hypotheses six and seven some of the results of the site-by-site analysis should be taken as indicative only, because the sample sizes are, in a few cases, too small for reliable testing. Hypothesis six was tested using both the WTP-PTPR and the standard-bidding PTPR perceived discount rate data. Results are presented in Table H6.

At sites 3 and 4, the null hypothesis is rejected for the WTP-PTPR results. At sites 1 and 2 the null hypothesis is accepted. For the standard bidding PTPR, z is significant at site 1, but adopters have a higher standard-bidding PTPR than non-adopters (the significantly different results for WTP PTPR showed adopters with a lower discount rate). Site 4 shows significant difference, but z is not significant at site 2 or site 3. Site 2 is the only site with outright ownership as the predominant tenure status. In summary, there appears to be a definite relationship between perceived discount rates and soil conservation adoption at some sites, depending on the measure of discount rates used. However, we must bear in mind that these results are indicative but not conclusive because of the small sample sizes.

Table H6 Summary of statistical testing for hypothesis six

Criterion	Site 1	Site 2	Site 3	Site 4
Mean WTP PTPR (%) adopters	55	49	20	33
s.d.	27	17	10	20
Number (n)	16	20	5	13
Mean WTP PTPR (%) Non-adopters	64	51	52	47
s.d.	21	14	37	19
Number (n)	19	19	35	22
Z	1.1	0.4	4.2*	2.1*
Mean std-bid. PTPR (%) adopters	61	44	35	38
s.d.	26	17	14	20
Number (n)	16	20	5	13
Mean std-bid. PTPR (%) Non-adopters	47	45	42	50
s.d.	20	16	19	15
Number (n)	19	19	35	22
Z	1.8*	0.2	1.0	1.9*

* Significant at least at the 95% level of confidence

The results of testing hypothesis seven are presented in Table H7. The results are statistically insignificant at the 95% confidence level at all sites. Therefore we can accept the hypothesis that there is no statistical difference between mean interest rates paid by adopters and non-adopters of soil conservation. Again, some of the sample sizes are too small to be more than indicative. However, the results are unsurprising since the respondents appear to have little choice about what interest rates they pay.

Overall, the data are very complex. Many farmers are cultivating several plots and each plot may be held under a different form of land tenure. The predominant land tenure criterion classifies blocks acording to the tenure system relating to more than 50 % of the farmer's holdings. In addition, the adopting or not adopting criterion is determined by the farmer's own perception of whether he or she is practising soil conservation. All discount rates and interest rates are nominal. It is assumed that with standard-bidding PTPR questions respondents are factoring inflation into their responses.

Table H7 Summary of statistical testing for hypothesis seven

Criterion	Site 1	Site 2	Site 3	Site 4
Mean interest rate (%) Adopters	116	192	151	413
s.d.	57	137	9	165
Number (n)	11	14	2	7
Mean interest rate (%) non-adopters	153	135	180	397
s.d.	65	76	95	246
Number (n)	15	11	16	15
Z	0.8	1.2	1.2	0.2

It is likely that a range of factors, some of which are unknown, are affecting soil conservation adoption. Land tenure, discount rate and income do appear to have an influence but other issues such as landlords' attitudes, peer group pressure, perceptions of the effectiveness of available technologies and cultural factors are likely to have an influence. There may even be a synergistic effect between these factors which it is not possible to demonstrate using statistical techniques and in the absence of further information.

Policy Proposals

As mentioned earlier, there is no single factor that can be clearly identified as influencing soil conservation adoption and sustainable land-use. However, it is concluded that land tenure, income, perceived discount rate and fair credit are inter-related variables that between them have a very strong influence. The importance of the fourth of these variables, the availability of fair credit, is strongly influenced by the significance of private time preference rates in long-term decision making. Farmers with high discount rates and numerous claims on their limited resources (claims such as feeding themselves and their families) cannot take long-term risks by investing in technologies with short-term costs and long-term benefits.

The significance of fair credit costs and conditions is that in most cases farmers would have to borrow to invest in conservation activities. But since many of them borrow money to buy food (a consumption activity they cannot defer), the cost of credit and the relationship of interest on debt to their own discount rate will affect their decision-making processes, their attitude to risk and their utility maximising activities. If interest rates are substantially higher than their already high discount rates, it is very unlikely that they will choose to invest in long-term projects. The survey results indicate that if not borrowing for food or medicine, farmers frequently are borrowing to avert disasters such as the loss of their crops or of their land. The upland farmers are operating at the margin and need to reduce the risk or effects of such marginal activity. In addition, the survey results indicate that neo-classical assumptions linking interest and discount rates, and predicating farmer decision making on the basis of unvarying personal discount rates, are not applicable to the market situation at the study sites. Given that much economic policy has been theory driven, it is time to examine farmer decision making in a different light, and to propose policy which reflects this.

Some basic findings of this research are as follows:

- interest rates on outstanding debt were not a surrogate for discount rates, for respondents at the study sites;
- private time preference rates are not constant at any given time for individuals in a range of hypothetical situations;
- perceived discount rates show variation with duration, magnitude, purpose and nature of the loan;
- higher willingness to pay does not necessarily indicate a greater subjective preference for that option;
- WTP-PTPR is higher, and closer to the observed interest rate on debt, when the respondent is proposing to borrow to avert severe loss or to attract substantial gain, with the highest WTP PTPR for averting a serious loss;
- the higher the median observed interest rate, the smaller the median loan value;
- mean and median perceived discount rates at all sites show much less variation than mean and median observed interest rates at all sites;
- poor landowners are more likely than poor tenants to adopt soil conservation practices, and poor landowners are as likely as relatively well-off tenants to adopt soil conservation practices.

- farmers adopting soil conservation practices generally have lower mean perceived discount rates than non-adopting farmers. Site medians for perceived discount rates are lower for adopting farmers than for non-adopting farmers, except at Site 1 where they are equal (see Table 8.3).

Altieri and Masera (1993), writing of Latin America, suggest that sustainable rural development policies must focus on the principal development priorities of the region. They suggest (p. 95) that the policies should address:

- reduction of poverty;
- adequate food supply and self-sufficiency;
- natural resource conservation; and
- empowerment of local communities and the effective participation
- of the rural poor in the development process.

These goals are relevant to the Philippines, but the question of how the goals might be addressed is of central importance.

Opschoor and van der Straaten (1993, p. 218) recognise the gap between theory and reality in policy formulation. They assert that nature and the environment have never been in a worse state, partly because of ... 'the existence of a dominant economic theory in which insufficient categories are found to adequately analyse and solve continental and global environmental problems of a fundamental character'. They propose that ... 'The way out can be found in the incorporation of natural resources into the economic theory itself' and that (p. 219) ... 'The use of the interest rate as is found on the market, without taking into account the ecological importance of the future, should be discussed in the theories of finance.'

One of the most important factors influencing the adoption of soil conservation practices in Leyte is the high perceived discount rates of individual farmers and their divergence from the even higher interest rates on outstanding debt.

An important policy initiative would be to offer subsidised credit, without extreme penalties, and subject to an income means test to farmers requiring assistance. There could be several categories of subsidy eligibility. Because landowners are generally more likely to adopt soil conservation practices, extra incentives should be offered to tenants. There should be some sort of scheme to elicit the involvement and support of landlords since it is likely that their attitudes will influence their tenants' adoption practices.[2] The incentives for soil conservation adoption could be jointly offered by the Philippines government and aid-agencies, or by aid agencies alone once the issue of determining the appropriate interest rates to offer has been resolved. In Table 8.3 median standard-bidding PTPR, the rate of soil conservation adoption, the median income and predominant land tenure status are shown.

A policy could also be developed for offering joint subsidised loans to co-operatives of farmers, to help address the externalities (and disincentives) due to uneven adoption of soil conservation causing external benefits to accrue to non-adopters, or external costs to accrue to adopters with non-adopting neighbours.

Table 8.3 Summary of factors influencing soil conservation adoption

Site	Median PTPR adopters (non-adopters)	Soil cons. adopt. rate (%)	Median site income (pesos)	Predominant tenure status
1	50 (50)	43	15,021	tenant
2	25 (50)	50	9396	outright owner
3	27 (50)	12	8349	tenant
4	38 (50)	42	6794	amortising owner

The government, which is in control of forestry licences and polices illegal logging, should also develop policies for controlling externalities from up-slope logging activites which may give disincentives to upland farmers in the path of runoff and landslides.

In addition, agrarian reform programs should be equitably implemented, with fair land purchase prices for amortising owners, to increase the number of farm owners. The owners are likely to need fewer incentives for sustainable land use adoption.

Another area for policy development is to educate landlords (particularly absentee landlords) about the long-term benefits of soil conservation adoption. We have no information about the role of landlords in sustainable landuse and this is an area in which research needs to be conducted. However, if landlords were aware of benefits from soil conservation they might contribute to its adoption, or be induced to provide their tenants with incentives.

It may also be possible to instigate a program like the Grameen Bank which was established to help poor rural women to obtain fair credit in Bangladesh (Varian, 1993, p. 621).[3] Without the bank, villagers only have access to a monopolist usurer (moneylender) and, as Varian (*ibid*) states:

> Such a local monopoly is especially pernicious in an underdeveloped country such as Bangladesh. At an interest rate of 150% there are many profitable projects that are not being undertaken by the peasants. Improved access to credit could lead to a major increase in investment, and a corresponding increase in the standard of living.

Notes

1. It is recognised that moneylenders do face high costs for loan management, and that the price of credit is probably increased by the transaction costs associated with extending a large number of small loans.

2. At site 1 it is possible that the landlord is already playing a role in soil conservation adoption, given the high degree of soil conservation adoption among tenants. In addition, at this site alone the mean perceived discount rate appears to be higher for non-adopters, while the median standard bidding PTPR is the same for adopters and non-adopters. At the other sites median adopters' PTPR is much lower.

3. In this scheme five villagers with individual projects co-operate to apply for loans. If the loan is given approval, two group members start with a loan and invest in their projects, if they are able to repay according to the loan schedule the next two members may borrow, and if they are successful, the last member (the group leader) obtains a loan. The bank makes about 475,000 loans a month and has a default rate of about 2%.

Chapter 9

Quo Vadis?
Summary and Conclusions

Introduction

A discussion about the development and importance of the concept of sustainability of resource use, and its global and theoretical relevance today, was presented in Chapter 1 of this book. The problem of soil degradation in the Philippines' uplands was then discussed in the context of sustainable development. Some tentative hypotheses about the reasons for non-adoption of known soil conservation practices were proposed. These related to the influence of land tenure on efficient resource use, the influence of the system of credit and loan arrangements on farmers' borrowing patterns, and the influence of cropping patterns on land degradation and farmer decision making. It was hypothesised that the overall constraint on farmer decision making was poverty and debt, and that these factors influenced most other issues which affected the day-to-day life of the farmers.

Following the introductory chapter, there was a brief assessment of the social and economic history of the Philippines in Chapter 2. That chapter set the scene for the rest of the research, describing settlement patterns in the Philippines, origins of the social and political hierarchy, the influences of Spanish and American colonisation, the development of the land tenure system, and the influence of protected markets on the establishment of export and subsistence cropping patterns.

The existing socio-economic and land-use systems have a strong foundation in the history of the Philippines and these in turn have had a profound effect on the ability of small upland farmers to make free choices about all aspects of their lives. Because the land-use system in the Philippines has changed from one of sustainable agricultural practices to one of severe land degradation and unsustainable agriculture in the last 300 years, the Philippines has provided a useful case study of the non-adoption of soil conservation practices and related factors.

Chapters 3 and 4 have covered geophysical and socio-cultural aspects of Leyte and the Philippines, as well as discussing the background and establishment of this project. That discussion included brief descriptions of the study sites. Development of the socio-economic survey, upon which the whole study was based, was also discussed. In Chapters 4, 6, 7 and 8 there was an evaluation of previous studies in this field, presented in logical order according to the particular piece of theory or discipline to which they belong. In Chapter 5 a summary of the relevant theory of interest and discounting was presented, in order to establish and clarify the theoretical assumptions within the discipline of economics, and to examine the range of perspectives of different practitioners in this field. Discounting theory underpins the analysis of borrowing and lending in the Philippines presented in this book. Questions were raised about the credit market and the influences upon it at the various study sites. It has not been possible to answer all the questions raised here. Nevertheless general knowledge about borrowing and lending activities and their influence on farmer decision making has been increased as a result of the analysis presented in chapters 6, 7 and 8.

Throughout the study care was taken to establish the reliability of the data and of the assumptions upon which the research was predicated. Wherever there was any doubt about reliability, a clear statement of the problem was presented.

This, the ultimate chapter of the book, serves to present a summary of all significant results with a statement of important findings.

Summary of Results

The results of this study have been discussed in detail in earlier chapters of the book. The socio-economic surveys have generated a great deal of information and one of the goals in the book has been to present the results which are most relevant to policy development, and to the reassessment of some underlying assumptions in standard economic theory. The research in this book shows that market interest rates paid by farmers at the study sites are different from their individual discount rates, and that great caution should be applied in using market interest rates as a shadow for private discount rates. These issues were discussed and analysed in Chapter 8, and they will be distilled here. A summary of borrowers, adopters and mean discount rates for each site is presented in Table 9.1.

Table 9.1 Soil conservation adoption, borrowing and discount rates (WTP)

Site	Percentage Borrowing	Percentage Claiming to adopt	Mean WTP PTPR (%) adopting	Mean WTP PTPR (%) non-adopting
1	67	43	55	64
2	67	50	49	51
3	47	12	20	52
4	62	42	33	47

The main results from the research presented here are summarised below:

- The interest rate actually paid on outstanding debt shows great variation between individuals within sites and from one site to another. The observed interest rates paid by individuals are very different from their perceived discount rates, generally being much higher, and should not be used as shadow for their discount rates. This was shown to be the case when relevant hypotheses were tested in Chapter 7.
- An individual's discount rate varies according to the nature of the loan, the duration of the loan and the purpose of the loan.
- Individuals demonstrate a much higher willingness to pay (WTP) for a loan when the purpose of the loan is to avert a severe financial loss than when the purpose of the loan is unspecified.
- Many of the respondents are borrowing money to buy food or medicine, or to pay for funerals, rather than for, or in addition to, borrowing for farm inputs.
- An individual with a higher income does not always have a lower discount rate than an individual with a low income. This was demonstrated when the mean standard- bidding PTPRs at all sites were compared. In addition, the mean PTPR at the wealthiest site was higher than the mean PTPR at the poorest site.
- The site with the lowest number of borrowers has the lowest rate of soil conservation adoption.
- Respondents adopting soil conservation measures generally have lower perceived discount rates than those not adopting soil conservation measures.

These findings are generally consistent with Parrilla's results using the same survey samples (Parrilla, 1992). She found that age, education, wealth, discount rate and income affected the rate of uptake of soil conservation practices. However, she assumed, in keeping with economic theory, that the observed interest rates were a shadow for the discount rates, which is shown to be an inaccurate assumption. Parrilla's assumptions about the effect of individual discount rates on soil conservation appear to be correct. She states (pp. 19-20):

> Individuals and the community, taken as a whole, may discount future income at different rates. In general poorer people, placing a necessarily high premium on present consumption, have high discount rates ...
>
> Very high discount rates will be associated with complete failure to invest for the future, and this will of course have a major impact on the adoption of relatively long term investments such as soil erosion control.

Further:

> ... there may be another economic constraint associated with the ability to pay for purchased inputs associated with the technology – i.e. planting and other necessary material. As with different discount rates, this may partly be overcome by increasing the availability of credit at reasonable rates, or through the ... use of subsidies.

Research and Policy Options

The research presented in this book suggests some immediate policy options, especially with respect to the links between poverty, interest rates, discount rates and decision making. These will be discussed in the next section of this chapter. However, it has also become clear that there are several unknown factors that may influence farmer decision making, long-term environmental management and sustainable resource use. Viable options should probably be proposed as social policy initiatives because, although individual farmer decision making may be interpreted as a disaggregated microeconomic issue, the regional and social benefits or costs of land management improvement, or the lack of it, are likely to be substantial.

One factor about which there is little information at the study sites is the attitude and influence of the tenants' landlords, which may have a significant impact on farmers' decisions about adopting soil conservation practices. It is possible that the great difference between adoption rates at the two predominantly tenanted sites might have been affected by the landlords' attitudes to their tenanted lands. This could be an issue for further study. Some factors that might be worth assessing include:

- Socio-economic profile of the landlords including:
    - age, wealth, income, landholdings, form of rent collection (share?), place of abode (absentee?), manner of obtaining the land (inherited?), debt profile, interest paid on debts, personal discount rate, and lending activities.
- Landlords' attitudes to the land including:
    - productive or political asset?
    - investment in perpetuity or short-term financial investment?
- Landlords' attitudes to tenants
- Landlords' attitudes to soil conservation and environmental management.

Another avenue of further study could be to conduct time-series surveys of the debt profiles, incomes, discount rates and soil conservation activities of respondents, to see whether any of these factors change over time, and whether and how they change according to changing personal circumstances. These results could be cross-referenced with the results of a landlord survey to see if there is an ongoing link between tenants' practices and landlords' attitudes.

An additional suggestion for further study, one that is more general, and perhaps equally applicable in both developed and less-developed nations, would be to instigate a broadly based regional analysis of the effects of soil conservation, or its absence, over a number of catchments, taking into account climatic events and their consequences such as floods and landslides, ecological change, land degradation status, human health, water quality sustainability, and socio-economic profiles, in the manner of 'ecosystem health' and catchment management strategy approaches to environmental and social policy.

In areas of the globe affected by ecosystem damage, it is becoming increasingly important to assess and understand the long-term tangible and intangible social and environmental costs of all types of degradation. This includes biodiversity loss, as well as the loss of arable lands and the diminution of potable water supplies. This point is important because there is a divergence between society's perception of individual responsibility and its perception of collective responsibility. In addition there is an inconsistency in the interpretation and application of discounting and other aspects of economic theory, which tend to apply economics to policy in the context of 'finance' or 'commerce' (Lumley, 1999).

For example, a standard procedure for policy development and project selection is to use a discounted cash flow analysis to choose the project with the highest net present value (NPV), which, as mentioned earlier, introduces a very strong bias against projects with long-term benefits and short-term costs. Large companies and governments frequently make decisions on the basis of which option has the highest financial NPV. This approach often leads to non-selection of projects which have intangiible (non-market) environmental and social benefits (or lower environmental and social costs). For example, clear-felling old-growth forests, such as the Philippine rainforests, will have a much higher NPV than establishing a forestry plantation elsewhere (which may even have a negative NPV). This manner of decision making is deemed to be economically rational by many policy makers, yet it will not account for externalities such as landslides and floods which lead to death, homelessness, and land and water degradation.

On the other hand, it is often suggested that individuals who do not adopt conservation practices which will accrue long-term benefits are not behaving rationally. Yet part of the reason for such behaviour is that unconsciously they appear to discount benefits to the present. If their personal discount rates are very high they are unlikely to adopt such practices (and nor, presumably, would a commercial company) (Lumley, 1997). There appear to be different expectations placed on marginal individuals using natural resources for profit, compared with large companies using natural resources for profit. To augment the impact of such an inconsistency, companies often expect to derive their profit from publicly owned natural resources, such as in timber operations the world over, while marginal farmers, particularly in less-developed nations, are expected to earn their living from small parcels of private land on which they are frequently share tenants.

Either way, the social costs of discounting behaviour may be extremely high. The point of conducting an all-encompassing social-benefit cost analysis, which accounts for all externalities and all non-quantifiable intangible effects (possibly descriptively), may be to make explicit the long-term inter- and intra-generational costs of such discounting behaviour, and thus to inform social policy for those who require identification of explicit costs. It is worth bearing in mind Schulze's comments (Schultze 1994, p. 199) in his critique of cost-benefit analyses and environmental policy, where he states:

Lest cost-benefit analyses do more harm than good, it is critical that their biases be given due consideration when results are interpreted ...

Failure to carefully weigh projected net present values against the less quantifiable and presumably less well understood impacts that were not included in the model not only risks imparting an errant conclusion with the masquerading precision of a quantitative result ... Some complications may be reduced, but others become hidden behind quantifable, discounted, often apparently precise values, with the consequence that the attempt to reduce complexity runs the risk, if the results are not carefully interpreted, of decreasing rather than increasing the likelihood that a sound analytical foundation will serve as the basis for policy debates.

Two other factors for which additional study might be used to inform policy concern the productivity effects of soil-conservation adoption, and the pay-back period on the investment in soil conservation technology required by potential adopters.

There is a range of possible practices which may be adopted to ameliorate land degradation, but there is often insufficient information about which of these is best under which circumstances, and about exactly what productivity gains will be achieved from the farmers' perspectives.

The magnitude of productivity effects at the study sites examined in this book is likely to be site-specific, as will be the long-term effects of the various options on land degradation. As mentioned earlier, a farmer's perception of the pay-back period for return on the investment in the technology is likely to affect his or her adoption activity. Since we do not have sufficient information on the financial effects of productivity improvements at the Philippines study sites it is not possible to estimate the exact pay-back periods concerned.

However, farmers will have their own assessments of likely pay-back periods, and their decision making will be influenced by their knowledge and perception of the subject. It is likely that required pay-back periods for many farmers will be very short (see Parrilla, 1992), possibly being in the realm of 100% payback for between one harvest cycle and one year. Since this is likely to have some influence on conservation adoption there is a need for further research on productivity gains and farmers' required pay-back periods.

Policy Perspectives

Access to fair credit has an impact on the rate of soil conservation adoption among upland farmers. Because of the long-term nature of the benefits of adoption, the farmers' own discount rates play an important role in adoption practices. The higher the interest rate on outstanding debt, the lower the likelihood of farmers adopting soil conservation. Farmers with lower discount rates generally have a higher rate of soil conservation adoption, and borrowers have a higher rate of soil conservation adoption than non-borrowers (despite many borrowings being spent on non-farm items like food). This indicates a link between interest rates, discount rates and soil conservation adoption. Similar links have been observed elsewhere (Godoy *et al*, 2001; Moseley, 2001).

In addition, there is a link, though less clearly defined, between land tenure status and soil conservation adoption. Generally land owners are more likely to adopt than tenants. However, given the difference in adoption rates between the sites where the farmers are predominantly tenants, there are also unknown factors involved, although it is probable that the attitude of the landlord has some influence, as discussed above. In the Philippines at least, the effect of land tenure or property rights needs more in-depth consideration given the historical context of settlement patterns and the development of farming. There the original land-use system was one of common property/usufruct followed by a type of feudalism, which then gave way to complete privatisation and alienation of the land (Seekins, 1984; Goodno, 1991; Lumley, 1998 (c)).

Aguilera-Klink (1994, p. 227) discusses the issue of environmental management and property rights generally, stating:

> ... having acknowledged the global interdependence and biological context that characterise human activity, it would appear that a new institutional framework must be put in place to cater adequately to this situation and to show clearly that property rights, whether they be private, public or common, are by no means absolute.

Aguilera-Klink (*ibid*) thus identifies a policy problem which arises from a possible misunderstanding of the classical economists' approach to common property leading to the 'tragedy of the commons' argument (Hardin, 1968). Current land-tenure systems around the world probably owe something to the 'all privatisation is good' manner of interpretation of Adam Smith's and other classical economists' work on market economics, their caveats about public responsibility and 'moral sentiments' going unheeded (Lumley, 2000). Land tenure is certainly an issue that needs consideration from a policy perspective at the four study sites. Unfortunately, there is not enough information here about the role of landlords to enable definitive proposals at this stage, although analysis of possible attitudes among landlords was presented in Chapter 8.

When proposing policy for achievement of specified goals, the cheaper the policy the more likely it is to be implemented. Education and information dissemination are usually good first steps in policy development, although in most circumstances a combination of approaches will constitute the most efficient and most effective policy.

At the study sites the farmers were nearly always aware of the full range of technologies available for soil conservation, as they were also aware of the benefits from soil conservation adoption. Yet many of these farmers (in fact, when all the sites are aggregated, the majority of these farmers) do not adopt soil conservation. This leads to a non-optimal allocation of resources, since the land is becoming rapidly degraded and is not nearly as productive as it should be. In addition, land degradation leads to the generation of externalities, some of which, particularly those which lead to loss of life, have very high social costs. In neo-classical economic theory externalities are deemed to be inefficient and their internalisation, in cases of 'market failure' to require government intervention, sometimes in the form of cash or in-kind subsidies to affected parties (Common, 1995; Hussen, 2000).

The non-adoption of soil conservation among farmers is not, then, because they are ignorant, but is influenced by the very high cost of credit, and the farmers' own high discount rates. However, it is possible that the landlords of tenanted lands are ignorant of the full costs of land degradation, or of the full benefits of soil conservation adoption, in which case education of landlords would be an appropriate policy issue.

In the study presented here the farmers' responses to willingness-to-pay questions concerning loss and gain indicate that they are extremely risk-averse and are operating at the margin, upland farmers being subject to great pressures from their day-to-day family and business commitments and from the unpredictability of the climate. They cannot afford to take even a small risk in borrowing funds for implementing soil conservation which they might not be able to repay. This is especially the case when default on loans attracts severe penalties, and benefits of the technology are in the long term yet incur up-front costs.

As discussed earlier, one obvious policy is to provide subsidised credit, since access to fair credit appears to be of such overriding importance in decision making, and farmers have very little choice about the rate of interest they can pay on outstanding debts (Ledesma, 1982). However, the manner in which subsidised credit can be provided needs careful consideration. One option would be to establish a Grameen-style bank (as discussed in Chapter 8) but this might require the presence of a wealthy philanthropist and a great deal of collective responsibility. That is not to say that this is not a feasible alternative, but since the credit will be very specifically goal-oriented (focussing on optimal resource allocation in neo-classical economics terms, or on sustainable resource use in ecological economics terms) a more narrowly specified form of subsidy might be appropriate. The amount and level of subsidy needs to be carefully considered, as does the potential subsidy provider and eligibility criteria for potential subsidy recipients.

A possible approach would be to provide a means-tested subsidy with the effective amount of the subsidy based on the private time preference rate of the recipient for the purpose of soil conservation. However, this may be a clumsy tool given that estimation of discount rates is a very time-consuming procedure. In addition, it would be necessary for subsidy recipients to understand the basis on which they were selected. This could lead to at least two potential problems. The first being that discount rate is a hard concept to comprehend and the second being that those who do comprehend it may overstate their rate in order to attract a larger subsidy.

A better approach might be to provide an interest-free loan over, say, a five-year period, for the specific purpose of introducing soil conservation technology on the farm, similar to the successful Cleaner Production Grants once offered by an independent government body, the Australian Centre for Cleaner Production, in Victoria, Australia (Lumley, 1994). One problem with the Victorian scheme was that the total amount of grant money offered was too small to have a significant impact on pollution at a regional level.

The amount of the suggested subsidised soil conservation loan would need to be based directly on the cost of the technology and should probably cover the entire cost of the technology with capital repayment commencing only after the payback period has ended (up to three years). All applicants would be required to present an implementation plan showing expected costs and returns of the investment, which would be carefully monitored over the life of the loan. Education programs for farm planning and technology selection and implementation could be provided to those who require it. For tenanted farms, landlord participation should be encouraged and in such cases it may be possible to seek assistance from the landlord in subsidy provision. For example, the landlord might be asked to provide conservation inputs.

The problem of loan default should not arise if the scheme is carefully administered and implemented. For example, following assessment of the farm plan the loan could be provided in-kind, with specific inputs for the soil conservation program being the loan capital. An alternative to a subsidy program with long-term loan repayment could be direct donation of the technology concerned to eligible farmers. However, a repayment schedule based on a successful payback period would provide a stronger incentive to farmers to make the technology work.

Another policy, which could be implemented in tandem with subsidised credit, and with landlord and farmer education, would be a fair agrarian reform program. Clearly land tenure, along with fair credit and farmers' private time preference rates, does influence a farmer's decision to adopt soil conservation. Unless there are some very enlightened landowners involved, only governments can instigate agrarian reform programs which allow tenant farmers to purchase, at a fair price, the land they farm. Unfortunately, governments the world over, including those of the Philippines, India, Peru, Ecuador and Thailand, have a very poor Agrarian Reform record, even when such reform has been legislated (Handelman, 1980; Ledesma, 1982; Rao, 1994). However, if governments understand the social and economic costs of land degradation, they may be more prepared to implement agrarian reform.

Whichever policies are selected for implementation, they should be both efficient and equitable (often a difficult coupling) and must have a beneficial effect on social and environmental sustainability. The policies proposed above, which would be implemented most effectively together, should me*et all* of these criteria and can be summarised under the headings education, provision of fair credit, agrarian reform and research. While this book has focussed on the Philippines, it is clear that these general policy headings are also important to poor farmers and land managers the world over (Ledesma, 1982; Kohli, 1987; Rao, 1994; Boyce, 1994; Martinez-Alier, 1995).

Summary and Conclusions

Major findings presented in this book are outlined as follows:

- Because farmers are a heterogeneous group, mean and median incomes vary greatly between sites, so any generalisations must be made in the light of these differences.
- Income equality varies between sites (there is a very skewed income distribution), and greater equality appears to influence higher conservation rates.
- Perceived discount rates at all sites are very high but surprisingly consistent given the differences in interest charged on debt, income and land-tenure status.

- Perceived discount rates are much lower than interest rates at all sites except when specifically relating to questions about averting large financial losses or attracting large financial gains. Then discount rates approximate interest rates more closely, but do not equal them.
- Soil conservation adoption rates vary across all sites but there is a link with perceived discount rates. Farmers who perceive themselves to be practising soil conservation generally have lower discount rates than farmers who do not have this perception.
- Land-tenure status (in combination with income and perceived discount rate) probably has a substantial influence on conservation practices.
- A large proportion of farmers' borrowing is to purchase non-farm-input, non-luxury consumption items such as food.
- Because of farmers' very high discount rates, low incomes and lack of access to fair credit, the only policy likely to strongly influence soil conservation activities is to supply subsidised or free credit specifically for the establishment of soil conservation.
- The true tangible and intangible long-term costs of upland degradation should be assessed in a social and environmental sustainability framework in order to determine a perspective for overall subsidy levels. Cost-benefit analysis is an inadequate economic tool for environmental evaluation but one that is generally accepted by decision makers. So long as every contingency, is accounted for, including the valuation of intangibles, cost-benefit analysis can provide some useful insights.
- If the questions of low income, high discount rates, land tenure, and soil erosion are not thoroughly addressed, land degradation in countries such as the Philippines and possibly throughout the world may become irreversible.
- The application of the assumptions in economic theory should be considered carefully, particularly when concerned with individual interest and discount rates, which equate willingness to pay with willingness to be compensated; those which always equate a higher willingness to pay with a greater preference for an option (i.e. people may not be able to dissociate reality from personal preference such as in WTP for in-kind versus cash credit); those which presume that all individuals have perfect freedom of choice and know their own interests best; and those which assume alienated lands are more efficient than common lands.

There is a combination of reasons for the continued existence, and even acceleration, of land degradation, which influence farmers' decisions to adopt soil conservation practices. A number of these have been addressed in this book and quite a lot of further research needs to be conducted in order to determine more clearly the process by which farmers make decisions. However, a number of factors that can be addressed by policy changes have been identified as influencing the farmer decision-making process. If the policy suggestions outlined in this chapter could be refined and implemented, a significant contribution might be made to improving environmental and social sustainability. If implemented at regional and national levels in developed and less-developed nations alike, such policies might have a positive global impact on sustainability.

Assumptions in market theory which state that the market will always allocate most efficiently, even with respect to the environment, need to be thoroughly analysed and reassessed. It is probable that such theory is not applicable to village economies in the Philippines and other less developed countries, because the assumptions are not consistent with reality. In fact it is arguable whether all the assumptions underlying economic theory (for example, perfect knowledge, perfect competition and no externalities) are concurrently consistent with reality anywhere.

It is clear that, in standard economic parlance, natural resource allocation in the Philippines uplands is by no means optimal for a range of reasons discussed in this book. The Philippines has been used as a case study to address the socio-economic and environmental aspects of land degradation, which is a significant and growing problem globally.

Pimental *et al* (1994, pp. 203-204) succinctly reiterate the importance of land degradation as an important environmental and social problem as follows:

> Rapid land degradation is a major threat to the sustainability of world food supply and affects most of the crop and pasture land throughout the world ... Estimates suggest that agricultural land degradation can be expected to depress food production between 15 % and 30 % over the next 25-year period, unless sound conservation practices are instituted now ...
>
> *Soil erosion is the single most serious cause of this degradation* [my emphasis], occurring at rates of 16 t / ha / year in the U.S. ... The major cause is the employment of poor agricultural practices that leave the soil without vegetative cover to protect against water and wind erosion. Soil loss is particularly distressing because it takes approximately 500 years to reform 2.5 cm (1 inch) of topsoil under normal agricultural conditions ...
>
> Each year because of land degradation primarily by erosion, about 15 million ha of new land must be found for agriculture. About 10 million ha is used to replace losses caused by land degradation. And an additional 5 million ha must be found to feed the 93 million humans added yearly to

the world population. This added agricultural land tends to come from clearing of vast forest areas ... The spread of agriculture accounts for about 80 % of the deforestation now occurring world wide.

Land degradation, and particularly soil erosion, is a very real threat to sustainable resource use on earth. Many of the respondents in this study are subsistence farmers and their borrowing activities already indicate that some cannot produce enough food to sustain themselves. The effects of cyclone Thelma, which killed 8,000 people in Leyte in 1991, and caused widespread landslides, flooding and infrastructure destruction, demonstrate what happens to denuded and degraded lands under the influence of cyclonic rainfall. The social costs are very high. The results presented in this book give a good indication of the factors which influence farmers' decision making with respect to soil conservation adoption. While this research has a seemingly narrow focus, based as it is on an island in the Philippines, it could be argued that, with respect to farmer decision-making, the results have a broad application. Early in this book (see O'Brien, 1987, in Chapter 1) work on soil conservation among farmers in Australia was discussed. Australian farmers would be motivated by similar influences to the farmers in the Philippines, with debt status, discount rate, uncertain income and overall risk being major considerations. In addition, with respect to other areas of environmental management such as industrial pollution, waste management and recycling, decision makers in small and large businesses would be making their choices about the adoption of environmentally sound technologies on a similar basis and predicated on similar assumptions about short-term and long-term financial and social returns.

It is hoped that this book can make a contribution to understanding the factors which influence soil erosion and sustainable agriculture in the Philippines and beyond, by using as a template this case study of Philippines farmers. The results should also be directly applicable to understanding factors which influence decision making about other environmental matters by farmers and small business managers. This book concerns global issues, and answers to questions about sustainable resource use can only be found by examining specific issues as a part of wider research. All such research can then contribute to the resolution of global environmental and social problems.

The issue of defining, applying and measuring sustainability is by no means resolved. As illustrated by a number of writers from a range of disciplines (for example Lumley, 1991; Palo and Lehto, 1996; Bowers, 1997; Elliott, 1998; Pearce, 1999; Howes, 2000; Phillis and Andriatiatsaholiniaina, 2001), the matter is still being argued and analysed.

However, with regard to policy determination, the environment and the future, it is clear, as many writers have pointed out, that while we are refining our theories and identifying sustainability indicators, we know what we ought and ought not to be doing with regard to environmental resource use and conservation. As Huxley (1944) stated more than half a century ago '... in the relatively simple affairs of nature, where we know quite well what is likely to happen, we immolate the future to the present'.

Bibliography

Aguilera-Klink, Federico (1994), 'Some Notes on the Misuse of Classic Writings in Economics on the Subject of Common Property', *Ecological Economics* Vol. 9, pp. 221-228.

Altieri, Miguel A., and Masera, Omar (1993), 'Sustainable Rural Development in Latin America: Building from the Bottom up', *Ecological Economics*, Vol. 7(2) pp. 93-121.

Anderson, John (1994), Personal Communication, La Trobe University, October.

Arce, Wilfredo F. and Ricardo G. Abad (1986), 'The Social Situation', in Bresnan, John (ed), *Crisis in the Philippines. Marcos and Beyond.* Princeton University Press, Princeton, New Jersey.

Archibugi, F. and Nijkamp, P. (eds) (1990), *Economy and Ecology: Towards Sustainable Development*, Kluwer Academic Publishers, Amsterdam.

Argy, Fred (1993), Presidential address to the 1993 Conference of Economists, Economic Society of Australia.

Arrow, Kenneth J. (1977), 'Criteria for Social Investment', in R. Dorfman and N. Dorfman, *Economics of the Environment, Selected Readings*, W.W. Norton and Co., New York and London.

Barbier, Edward, Burgess, Joanne C., and Folke, Carl (1994), *Paradise Lost? The Ecological Economics of Biodiversity*, Earthscan Publications, London.

Barbier, Edward (1991), 'Environmental Dedradation in the Third World', in Pearce, David (ed), *Blueprint 2. Greening the World Economy*, Earthscan Publications, London.

Belsky, J.M. (1984), 'Stratification Among Migrant Hillside Farmers and Some Implications for Agro-Forestry Programs: A Case Study in Leyte, Philippines', Unpublished Masters thesis, Cornell University, 175pp.

Binoya, Saloma (1987), 'Synthetic Profile of Barangay San Vicente in Bontoc', unpublished field report, ViSCA.

Binoya, Saloma (1988), 'The Socio-Economic Constraints on the Adoption of Improved Cropping Methods by Upland Farmers in Leyte, Philippines: Site 1 report, San Vicente, Bontoc, Southern Leyte', paper presented at the project conference, Visayas State College of Agriculture, Baybay, Leyte, September.

Black, A., Duff, J., Saggers, S. and Baines, P. (2000), *Rural Communities and Rural Social Issues: Priorities for Research*, RIRDC, Canberra.

Blaikie, Piers (1985), *The Political Econonmy of Soil Erosion in Developing Countries*, Longman, New York.

Blamey, R., Common, M. and Quiggan, J. (1995), 'Respondents to Contingent Valuation Surveys: Consumers or Citizens?', *Australian Journal of Agricultural Economics*, Vol. 39, pp. 263-288.

Blamey, R., Bennett, J.W., and Morrison, M.D. (1999), 'Yea-Saying in Contingent Valuation Surveys', *Land Economics*, Vol. 2, 126-141.

Bowers, John (1997), *Sustainability and Economics. An Alternative Text*, Longman, Harlow.

Boyce, J.K. (1994), 'Inequality as a Cause of Environmental Degradation', Ecological Economics, Vol. 11, pp. 169-179.

Brookes, Lance E. (1983), 'Upland food crop systems. Hillside Farming/Agroforestry', in ZDSDP, *Developing Tropical Uplands*, Zamboanga del Sur Development Project, ACIL International Pty. Ltd.

Bunge, Frederica, M. (ed) (1984), *Philippines. A Country Study*, the American University, Washington DC.

Buttel, Frederick H. (1998) 'Some Observations on States, World Orders and the Politics of Sustainability', *Organisation and Society*, Vol. 11 (3), pp. 261-286.

Cabrido. Candido. A. Jr. (1981). 'National Soil Erosion Control Management Project and Action Programs' (NASECOMPAP), Monograph. National Environmental Protection Council, Quezon City.

Cabrido, Candido, A. Jr. (1985), 'An Assessment of National Soil Erosion Control Management Programs in the Philippines', in E.T. Craswell, J.V. Remenyi and L.G. Nallana, *Soil erosion management*, ACIAR Proceedings Series No. 6.

Celestino, Andres F. (1985), 'Farming Systems Approach to Soil Erosion Control and Management', in E.T. Craswell, J.V. Remenyi and L.G. Nallana (eds) *Soil erosion management*, ACIAR Proceedings Series No. 6.

Center for Research and Communication (CRC) (1988), *Philippines Agribusiness Factbook and Directory 1987-1988*, Southeast Asian Science Foundation Inc, Manila.

Chick, V. (1985), 'Keynesians, Monetarists and Keynes: The End of the Debate - or a Beginning?', in Arestis, Philip and Skouras, Thanos (eds), *Post Keynesian Economic Theory. A Challenge to Neo-Classical Economics*, Wheatsheaf Books, Sussex.

Chivaura-Mususa, C., Campbell, B. and Kenyon, W. (2000), 'The value of Mature Trees in Arable Fields in the Smallholder Sector, Zimbabwe', *Ecological Economics*, Vol. 33, pp. 395-400.

Christodoulou, Demetrios (1990), *The Unpromised Land. Agrarian Reform and Conflict Worldwide*, Zed Books Ltd, London and New Jersey.

Citizens Disaster Rehabilitation Center (1991), '"Uring" Disaster Blamed on Denuded Forest', CDRC Bulletin Update No. 3, 9 November, CDRC, Quezon City, Philippines 2pp.

Common, Michael (1995), *Sustainability and Policy: Limits to Economics*, Cambridge University Press, Cambridge.

Commonwealth of Australia (1982), *Survey of Major Western Pacific Economies*, Department of Trade, Australian Government Publishing Service, Canberra.

Commonwealth of Australia (1983), *Survey of Major Western Pacific Economies*, Department of Trade, Australian Government Publishing Service, Canberra.

Commonwealth of Australia (1984), *Survey of Major Western Pacific Economies*, Department of Trade, Australian Government Publishing Service, Canberra.

Commonwealth of Australia (1986), *Survey of Major Western Pacific Economies*, Department of Trade, Australian Government Publishing Service, Canberra.

Commonwealth of Australia (1990), *Ecologically Sustainable Development. A Commonwealth Discussion Paper*, Australian Government Publishing Service, Canberra.

Commonwealth of Australia (1992), *Australian Interim National Strategy for Ecologically Sustainable Development*, Australian Government Publishing Service, Canberra.

Commonwealth of Australia (1994), *Summary Report on the Implementation of the National Strategy for Ecologically Sustainable Development*, Ecologically Sustainable Development Steering Committee, Australian Government Publishing Office, Canberra.

Commonwealth of Australia (2000), The Environment Protection and Biodiversity Conservation Act 1999, Australian Government Publishing Service, Canberra.

Community Aid Abroad (CAA) (1984), *The Third World war. The Philippines Front*, CAA, Fitzroy, Victoria.

Concepcion, Roger N. (1983), 'Soil and Water Conservation Strategies in Hilly Lands', in *Proceedings: Hilly Land Development Workshop*, Ecotech Center, Cebu City, March.

Corpuz, I.T. (1986), 'Soil Erosion Process', in *Soil Erosion Control Management. Proceedings of the first regional short term course on soil erosion control management of field trainers. Baguio City. Philippines*. Cited in Santa Maria Maniego (1986), Unpublished Masters thesis, University of the Philippines at Los Banos.

Cramb, R.A. (ed) (2000), *Soil Conservation Technologies for Smallholder Farming Systems in the Philippine Uplands: A Socioeconomic Evaluation*, Australian Centre for International Agricultural Research, Canberra.

Cramb, R.A. and Saguiguit, G.C. (2000), 'Introduction', in Cramb, R.A. (ed) (2000), *Soil Conservation Technologies for Smallerholder Farming Systems in the Philippine Uplands: A Socioeconomic Evaluation*, Australian Centre for International Agricultural Research, Canberra.

Dalton, John B. (1986), 'Developing Smallholder Rainfed Agriculture in the Philippines Uplands', Draft paper, June.

Daly, H.E. (1990), 'Toward Some Operational Principles of Sustainable Development'. *Ecological Economics*, Vol. 2, pp. 1-6.

D'Arge, R.C. (1990), 'The Promise of Ecological Economics', *J. Soil and Water Conservation*.

Dasgupta, A.K. and Pearce, D.W. (1972), *Cost Benefit Analysis: Theory and Practice*, Macmillan, London.

Davis, Leonard (1987), *The Philippines. People, Poverty and Politics*, Macmillan, London.

De Jesus-Viardo, A. (ed) (1984), 'The educational dilemma of women in Asia, 1969', cited in CAA, *The Third World War. The Philippines Front*, Community Aid Abroad, Fitzroy.

Dragun, A.K. and Jakobsson, K.M. (1992), 'Sustainable Institutions', paper presented to the Second International Society for Ecological Economics Conference, Stockholm, August.

Echstein, Otto (1958), *Water Resource Development: The Economics of Project Evaluation*, Harvard University Press.

Edgerton, R.K. (1984), 'The Society and its Environment', in Bunge, Frederica, M. (ed), *Philippines, A Country Study*, the American University, Washington.

Ekins, Paul and Max-Neef, Manfred (eds) (1992), *Real Life Economics*, Routledge, London and New York.

Elliott, Lorraine (1998), *The Global Politics of the Environment*, Macmillan, Basingstoke and London.

Farris, N.F. (1948), 'Soil erosion experiment at State College, Pa. Agri. Expt. Sta. Bul', cited in Santa Maria Maniego, unpublished masters thesis, University of the Philippines at Los Banos, 1986.

Fisher, Irving (1965), *The Theory of Interest*, Augustus M. Kelly, New York.

Floro, Sagrario and Yotopoulos Pan A. (1991), *Informal Credit Markets and the New Institutional Economics, The Case of Philippine Agriculture*, Westview Press, Boulder.

Foreman, John (1899), *The Philippine Islands. A Political, Geographical, Social and Commercial History of the Philippine Archipelago*, Charles Scribner's Sons, New York.

Fox. Robert B. (1970). *The Tabon Caves. Archaeological Explorations and Excavations on Palawan Island, Philippines*, National Museum, Manila.

Friend, A.M. and Rapport, D.J. (1991), 'Evolution of macro-information systems for sustainable development', *Ecological Economics*, 3(1), pp. 59-76.

Friend, Theodore (1986), 'Philippines - American Tensions in History', in Bresnan, John, *Crisis in the Philippines, The Marcos era and beyond*, Princeton University Press, Princeton, New Jersey.

Fullard, H. and Darby, H.C. (eds) (1979), *The Australian University Atlas*, George Philip and Son, London.

Galbraith, Kenneth John (1987), *A History of Economics. The Past as the Present*, Hamish Hamilton, London.

Godoy, R., Kirby, K. and Wilkie, D. (2001), 'Tenure Security, Private Time Preference, and use of natural Resources among Lowland Bolivian Amerindians', *Ecological Economics*, Vol. 38 (1), pp. 105-118.

Goodno, James B. (1991), *The Philippines. Land of Broken Promises.* Zed Books. London and New Jersey.

Government of Victoria (1987), *Protecting the Environment. A Conservation Strategy for Victoria*, Victorian Government Printing Service, Melbourne.

Grigalunas, T., Opaluch, J.J. and Luo, M. (2001), 'The Economic Costs to Fisheries from Marine Sediment Disposal: Case Study of Providence, RI, USA', *Ecological Economics*, Vol. 38 (1), pp. 47-58.

Gruen, Lori and Jamieson, Dale (1994), *Reflecting on Nature. Readings in Environmental Philosophy*, Oxford University Press, New York and Oxford.

Guo, Z., Xiao, X. and Zheng, Y. (2001), 'Ecosystem Functions, Services, and their Values - A Case Study in Xingshan County of China', *Ecological Economics*, Vol. 38 (1), pp. 141-154.

Handelman, Howard (ed) (1980), *The Politics of Agrarian Change in Asia and Latin America*, Indiana University Press, Bloomington.

Hardin, G. (1968), 'The Tragedy of the Commons', *Science*, Vol. 162, pp. 1243-1248.

Hernandez, Carolina G. (1986), 'Reconstituting the Political Order', in Bresnan, John, *Crisis in the Philippines. The Marcos era and beyond*, Princeton University Press, New Jersey.

Horne, P. (1996), 'Deforestation as an Environmental-Economic Problem in the Philippines', in Palo, Matti and Mery, Gerardo (eds) (1996), *Sustainable Forestry Challenges for Developing Countries*, Environmental Science and Technology Library, Kluwer Academic Publishers, Dordrecht, Boston and London.

Howes, Michael (2000), 'A Brief History of Commonwealth Sustainable Development Discourse', *Policy, Organisation and Society*, Vol. 19(1), pp. 65-85.

Hueting, R. (1990), 'The Brundtland Report: A Matter of Conflicting Goals', *Ecological Economics*, Vol. 2, pp. 109-117.

Hurley F.T., Fitzgerald, B.C., Harvey, J.T. and Oppenheim, P.P. (1987), 'The Problems of Change, a Study of the Decision Making Processes of Victorian Graingrowers', Ballarat College of Advanced Education, Ballarat.

Hussen, Ahmed M. (2000), *Principles of Environmental Economics. Economics, Ecology and Public Policy*, Routledge, London and New York.

Hussan, S. (1998), Letter to the Editor, *Ecological Economics*, Vol. 24, pp. 89-101.

Huxley, Aldous (1944), *Time Must Have a Stop*, Chatto and Windus, London; Reprinted, Flamingo, London (1994, p. 266).

International Monetary Fund (IMF) (1991), *World Economic and Financial Surveys World Economic Outlook*, IMF, Washington DC.

Jacobs, Michael (1993), *Environmental Economics, Sustainable Development and Successful Economies*, Resource Assessment Commission, Australia, Occasional Paper Number 4, April.

Japan Environmental Council (2000), *The State of the Environment in Asia 1999/2000*, Springer-Verlag, Tokyo.

Jarasaya, Sisera (1994), Personal Communication, Department of Economics, La Trobe University, Australia.

JCWARFASC (1998), *Towards a Regional Forest Agreement for the South-West Forest Region of Western Australia*, Joint Commonwealth and Western Australian Regional Forest Agreement Steering Committee, Commonwealth of Australia and Western Australian Governments, Canberra and Perth.

Jimenez, Pilar. R. and Francisco, Josepha S. (1984), *The Rural Poor in Leyte: A Social and Historical Profile*, Research Centre, De la Salle University, Philippines.

Kennedy, Alison, Gillen, Jake, Keetch, Bob, Creaser, Colin, and the Mutitjulu Community (2001), 'Gully Erosion Conrtol at Kantju Gorge, Uluru-Kata Tjuta National Park, Central Australia', *Ecological Management and Restoration*, Vol. 2 (1), pp. 17-27.

KEPAS (1985), 'The Critical Uplands of Eastern Java. An Agro-Ecosystems Analysis, The Research Group on Agro-Ecosystems, Kelompak Penelitian Agro-Ekosistem (KEPAS).

Kidron, Michael and Segal, Ronald (1995), *The State of the World Atlas*, Penguin, Harmondsworth.

Knetsch, Jack L. (1993), 'Environmental Valuation: Some Practical Problems of Wrong Questions and Misleading Answers', Resource Assessment Commission (Australia), Occasional Publication Number 5, September.

Kohli, Atul (1987), *The State and Poverty in India*, Cambridge University Press, Cambridge.

Krutilla, John V. and Fisher, Anthony C. (1978), *The Economics of Natural Environments. Studies in the Valuation of Commodity and Amenity Resources*, Resources for the Future, Baltimore.

Kula, Ehrun (1988), *The Economics of Forestry. Modern Theory and Practice*, Croom Helm, London and Sydney.

Lande, Carl H. (1986), 'The Political Crisis', in Bresnan, John, *Crisis in the Philippines. The Marcos Era and Beyond*, Princeton University Press, New Jersey.

Lankester, E. Ray (2000), 'The Effacement of Nature by Man', *Organization and Environment*, Vol. 13, 2: 236-239, reprinted from Lankester, E. Ray (1913), *Science from an Easy Chair: A Second Series*, Henry Holt and Company, New York.

Layard, Richard (1980), *Cost-Benefit Analysis*, Penguin, Harmondsworth.

Ledesma, Antonio (1982), *Landless Workers and Rice Farmers: Peasant Subclasses under Agrarian Reform in Two Philippine Villages*, International Rice Research Institute (IRRI), Los Banos.

Librero, Aida R. (1983), 'Bibliography of Socio-Economic Studies on Hilly Lands in the Philippines', in *Proceedings*, Hilly Land Development Workshop. Ecotech Center, Cebu City, March.

Librero, Aida R. (1985), 'Socio-Economic Considerations in a Soil Erosion Management Program: Case Study of Two Provinces in the Philippines', in E.T. Craswell, J.V. Remenyi and L.G. Nallana, *Soil Erosion Management*, ACIAR Proceedings Series no. 6.

Liebig, J. Von (1859), *Letters on Modern Agriculture*, Wiley, New York. Cited in Mayumo, K., 'Temporary Emancipation from the Land: from the Industrial Revolution to the Present Time', *Ecological Economics*, Vol. 4(1), pp. 35-56, 1991.

Lightfoot, C., Quero, F.V., and Villaneuva, M.R. (1985), *Review of Research Methods and Findings*, FSDP-EV report no. 33, FSDP-EV, Tacloban City, Philippines.

Loch, R.J. (1985), *Soil Erosion in the Philippine Uplands. Observations of the Problem. Recommendations for Research*, Queensland Department of Primary Industries, Brisbane.

Lumley, Sarah (1983), 'The Economic Implications of Dryland Salting and Associated Stream Salinity in North-Central Victoria', Unpublished Masters thesis, La Trobe University, Victoria.

Lumley, Sarah (1988), 'Social Time Preference Rate', Project Evaluation Seminar, Session 3, Unpublished Paper, Department of Conservation, Forests and Lands, Victoria.

Lumley, Sarah (1991), 'An Economic Perception of Sustainable Development', Draft Discussion Paper, Department of Conservation and Environment, Melbourne.

Lumley, Sarah (1994), 'Incentives and Disincentives. A Case Study: The Adoption of Clean Technologies by Small Companies under the Cleaner Production Program', Draft Paper, Environment Protection Authority (EPA), Victoria, October.

Lumley, Sarah (1995), 'Farmers' Willingness and Ability to Pay for Credit', in Menz, K.M., O'Brien, G.C. and Parrilla, L.S. (eds) *Report of a Project to Determine the Socio-Economic Constraints on the Adoption of Improved Cropping Methods by Upland Farmers in Leyte, Philippines. A Profile of Upland Farming Systems on Sloping Land in the Philippines*, ACIAR Working Paper No. 46. Australian Centre for International Agricultural Research, Canberra. ISBN 186 320 153X, October.

Lumley, Sarah (1997), 'The Environment and the Ethics of Discounting: An Empirical Analysis', *Ecological Economics*, Vol. 20(1), pp. 71-82.

Lumley, Sarah (1998 (a)), 'Cost - Benefit Analysis, Ethics and the Natural Environment', *Indian Journal of Applied Economics*, Vol. 7 (1), pp. 95-113.

Lumley, Sarah (1998 (b)), 'The Theory and Application of Cost-Benefit Analysis', *Ecological Economics*, Vol. 20(1), pp. 71-82.

Lumley, Sarah (1998 (c)), 'Ferdinand Magellan (Fernao de Magalhaes) 1480-1521', in Armstrong, P.H. and Martin, G.J. (eds), *Geographers: Biobibliographical Studies*, Vol. 18, for The Commission on the History of Geographical Thought of the International Geographical Union, and the International Union of the History of the Philosophy of Science, Mansell, London and Washington.

Lumley, Sarah (1999), 'Interpreting Economics, Rhetoric and Sustainable Development: Some Implications for Policy Determination', *Australian Geographer*, Vol. 30 (1), pp. 35-39.

Lumley, Sarah (2000), 'A Tale of Bad Times; or Ecological Economy, Sustainable Development and Harriet Martineau', *Australasian Victorian Studies Journal*, Vol. 6, pp. 60-67.

Lumley, Sarah and Stent, Bill (1989), *The Socio-Economic Constraints on the Adoption of Improved Cropping Methods by Upland Farmers in Leyte, Philippines: An Interim Report*, Department of Conservation, Forests and

Lands, Victoria, Economics Unit Discussion Paper no. 50, Melbourne, 1989 Government of Victoria (29 pp).

Lumley, Sarah and Hercock, Marion (2001), 'Locational and Socio-Economic Variation in Public Perceptions of Economics and the Environment', *GeoJournal*, Vol. 51, pp. 235-244.

Lumley, Sarah, Hercock, Marion and Bryant, Joe (2001), 'Society, Economics and the Environment: A Public Perspective', *Ecological Management and Restoration*, Vol. 2 (1), pp. 72-74

Lyall, Kimina (2001), 'Bin Laden is Funding Abu Sayyaf', *The Australian*, Canberra, 19 September.

Manila Chronicle (1991), Editorial, 'Gov't must react with firm policy, not panic', Col. 1, paras 4, 5, 15 November.

Manila Times (1991), '"Uring" leaves 3,000 dead in Visayas', Col 5, paras, 4, 7, p. A6. 7 November.

Martinez-Alier, Juan (1990), *Ecological Economics. Energy, Environment and Society*, Blackwell, Oxford.

Martinez-Alier, Juan (1995), 'The Environment as a Luxury Good or "Too Poor to be Green"?', *Ecological Economics*, Vol. 13, pp. 1-10.

Mayumi, Kozo (1991), 'Temporary Emancipation from the Land: from the Industrial Revolution to the Present Time', *Ecological Economics*, Vol. 4 (1), pp. 35-56.

McCoy, A.W. and de Jesus, E.C. (eds) (1982), *Philippine Social History. Global Trade, Local Transformations*, Ateneo de Manila University Press, and George Allen and Unwin, Sydney.

McKenna, Thomas M. (1998), *Muslim Rulers and Rebels. Everyday Politics and Armed Separatism in the Southern Philippines*, University of California Press, Berkeley, Los Angeles, London.

McManus, P. (1996), 'Contested Terrains: Politics, Stories and Discourses of Sustainability', *Environmental Politics*, Vol. 5, pp. 48-73.

Mishan, E.J. (1975), *Cost-benefit analysis*, George Allen and Unwin, London.

Moseley, William G. (2001), 'African Evidence on the Relation of Poverty, Time Preference and the Environment', *Ecological Economics*, Vol. 38, pp. 317-326.

Mueller, Ute, Schilizzi, Steven and Tuyet, Tran (1999), 'The Dynamics of Phase Farming in Dryland Salinity Abatement', *The Australian Journal of Agricultural and Resource Economics*, Vol. 43 (1), pp. 51-68.

National Economic and Development Agency (NEDA) (1985), *1984 Regional Socio-Economic Profile and Development Report*, NEDA, Leyte.

Noble, Lela Garner (1986), 'Politics in the Marcos era', in Bresnan, John, *Crisis in the Philippines. The Marcos Era and Beyond*, Princeton University Press, New Jersey.

O'Brien, B. (1987), 'A Study of Farmer Decision Making - Implications for Policy Makers', Land Protection Division, Department of Conservation, Forests and Lands, Melbourne.

O'Brien, G.C. (1989), 'The Socio-Economic Constraints on the Adoption of Improved Cropping Methods by Upland Farmers in Leyte, Philippines', Coding Guides, Working paper number 3, ACIAR project number 8541, Australian Centre for International Agricultural Research, Canberra.

O'Brien, G.C. (1991), 'Economic aspects of sustainable land management: regional experience', in IBSRAM, *Evaluation for sustainable land management in the developing world*, Technical Papers, proceedings no. 12, Volume II. September.

Opschoor, Hans, and van der Straaten, Jan (1993), 'Sustainable Development: An Institutional Approach', *Ecological Economics*, Vol. 7(3), pp. 203-222.

Pacardo, E.P. (1983), *The Effect of Corn/Ipil-Ipil Cropping System on Productivity and Stability of Upland Agroecosystem*, UPLB/PCARRD Annual Report.

Palo, Matti and Mery, Gerardo (eds) (1996), *Sustainable Forestry Challenges for Developing Countries*, Environmental Science and Technology Library, Kluwer Academic Publishers, Dordrecht, Boston and London.

Palo, Matti and Lehto, E. (1996), 'Tropical Asian Deforestation and Sustainability Prospects', in Palo, Matti and Mery, Gerardo (eds) (1996), *Sustainable Forestry Challenges for Developing Countries*, Environmental Science and Technology Library, Kluwer Academic Publishers, Dordrecht, Boston and London.

Pannell, David J. (2001), 'Dryland salinity: Inevitable, Inequitable, Intractable?', Presidential Address, 45th Annual Conference of the Australian Agricultural and Resource Economics Society, Adelaide, January.

Parrilla, L.S. (1992), 'Socio-Economic Constraints on the Adoption of Soil Conservation Practices by Upland Farmers in Leyte, Philippines', unpublished PhD thesis, La Trobe University, Victoria, August (277pp).

Parrilla, L.S., O'Brien, G.C., and Quilkey, J.J. (1991), 'Cluster Analysis of Upland Farming Practices in the Philippines', paper presented to the 35th Annual Conference of the Australian Agricultural Economics Society, Armidale, NSW, February.

Parrilla, L.S., O'Brien, G.C., Stent, W.R. and Villaneuva, C.D. (1988), 'The socio-economic constraints on the adoption of improved cropping methods by upland farmers in Leyte, Philippines: Research methodology', paper presented at Project Conference, Baybay, Leyte, September.

Pearce, D.W. (1971), *Cost-Benefit Analysis*, Macmillan, London.

Pearce, D.W. (1989), 'An Economic Perspective on Sustainable Development', *Development*, Vol 2 (3).

Pearce, D.W. (1999), *Economics and environment. Essays on ecological economics and sustainable development*, Edward Elgar, Aldershot, 1999.

Pearce, D.W. and Turner, R.K. (1990), *Economics of Natural Resources and the Environment*, Harvester Wheatsheaf, London.

Pearce, D.W., Barbier E. and Markandya, A. (1990), *Sustainable Development. Economics and Environment in the Third World*, Earthscan Publications, London (217pp).

Pearce, D.W., Markandya, A. and Barbier, E. (1989) *Blueprint for a Green Economy*, Earthscan Publications, London.

Pearce, Jenny (1986), *Promised Land. Peasant Rebellion in Chalatenango, El Salvador*, Latin American Bureau, London.

Pender, J.L. and Walker, T.S. (1991), *Experimental Measurement of Time Preference in Rural India*, ICRISAT. Resource Management Program Economics Group. Progress Report No. 97, Pantachery Andrha Pradesh, India, ICRISAT, February.

Perna, Fernando and Santos, Vitor (1998), 'The Free Riding Behaviour in Culatra Island Case Study: Detection and Correction', *Indian Journal of Applied Economics*, Vol. 7 (2), 269-290.

Philippine Daily Enquirer (1991), 'Leyte death toll may pass 7,000 mark', Col. 3, p. 1. 10 November.

Philippine Daily Enquirer (1991), 'Editorial. Fingering fingerlings', Col. 1, paras 2, 3. 13 November.

Philippines Daily Globe (1991), 8 November.

Philippines Daily Globe (1991), Editorial. 'Still to blame', 13 November.

Phillis, Y.A. and Andriatiatsaholiniaina, L.A. (2001), 'Sustainability: an Ill Defined Concept and its Assessment using Fuzzy Logic', *Ecological Economics*, Vol. 37 (3), pp. 453-457.

Pimental, David, Herdendorf, M., Eisenfeld, S., Olander, L., Carroquino, M., Corson, C., Mcdade, J., Chung, Y., Cannon, W., Roberts, J., Bluman, L. and Gregg, J. (1994), 'Achieving a Secure Energy Future: Environmental and Economic Issues', *Ecological Economics*, Vol. 9, pp. 201-219.

Pinches, Michael and Brown, Ian (2002), Forthcoming article, *Far East and Australasia*.

Race, D. and Robins, L. (1998), *Farm Forestry in Australia: A Research Update*, Rural Industries Research and Development Corporation, Canberra.

Rady, Graham (1990), 'Technologies for Sustainable Agriculture on Marginal Uplands in Southeast Asia: An AIDAB Perspective', in G. Blair and R. Lefroy, *Technologies for Sustainable Agriculture on Marginal Uplands in Southeast Asia*, ACIAR Proceedings Series no. 33.

Rao, C.H. Hanumantha (1994), *Agricultural Growth, Rural Poverty and Environmental Degradation in India*, Oxford University Press, Delhi.

Rao, M.S. (1972), 'Soil conservation of India', ICAR, New Delhi, cited in Santa Maria Maniego, unpublished Masters thesis, University of the Philippines at Los Banos, 1986.

Regional Development Council (RDC) (1985), *Updated Eastern Visayas Region Development Plan 1984 to 1987*, Government of the Republic of the Philippines, Manila.

Reoma, Bonifacio (1987), 'Synthetic Profile of Barangay Pomponan of Baybay, Leyte', unpublished field report, Visayan State College of Agriculture, Baybay, Leyte.

Reoma, Bonifacio (1988), 'Field Survey Notes from Pomponan', unpublished report, Visayan State College of Agriculture, Baybay, Leyte.

Reoma, Bonifacio (1988), 'The Socio-Economic Constraints on the Adoption of Improved Cropping Methods by Upland Farmers in Leyte, Philippines, An Overview of site 2, Pomponan, Baybay, Leyte', Paper presented at the project conference, Visayas State College of Agriculture, Baybay, Leyte, September.

Resource Assessment Commission (RAC) (1992), *Methods for Analysing Development and Conservation Issues: the Resource Assessment Commission's Experience*, Research Paper Number 7, RAC, Canberra, December.

Robinson, Joan (1952), *The Rate of Interest and Other Essays*, Macmillan, London.

Rodriguez, Filemon C. (1985), *The Marcos Regime. Rape of the Nation*, Vantage Press, New York.

Saastamionen, O. (1996), 'Change and Continuity in the Philippine Forest Policy', in Palo, Matti and Mery, Gerardo (eds) (1996), *Sustainable Forestry Challenges for Developing Countries*, Environmental Science and Technology Library, Kluwer Academic Publishers, Dordrecht, Boston and London.

Sajise, Percy E. and Ganapin, Belfin J. Jr. (1990), 'An Overview of Upland Development in the Philippines', in G. Blair and R. Lefroy, *Technologies for Sustainable Agriculture on Marginal Uplands in Southeast Asia*, ACIAR Proceedings Series, No. 33.

Sakar, A. and McKillop, W. (1991), 'Economic Efficiency and Ecological Considerations in Resource Development Policies. A case study', *Ecological Economics*, Vol. 4 (1).

Santa Maria Maniego, Simeon (1986), 'Evaluation of *Ipil-Ipil* Buffer Strips as a Soil Conservation Technique in Sitio Barambang, Cagnukot, Villaba, Leyte', unpublished masters thesis, University of the Philippines, Los Banos.

Schulze, Peter C. (1994), 'Cost-Benefit Analyses and Environmental Policy', *Ecological Economics*, Vol. 9, pp. 197-199.

Seagrave, Sterling (1988), *The Marcos Dynasty*, Macmillan, London.

Seekins, D.M. (1984), 'Historical Setting', in Bunge, F.M. (ed), *Philippines. A Country Study*, The American University, Washington DC.

Sharma, R.A., McGregor, M.J. and Blyth, J.F. (1991), 'The Social Discount Rate for Land Use Projects in India', *Journal of Agricultural Economics*, Vol. 42 (1).

Solheim, Willhelm, G. (1974), 'Potsherds and Potholes: Philippines Archaeology in 1974', in Center for Southeast Asian Studies, Special report no. 10, *Philippine Studies: Geography, Archaeology, Psychology and Literature. Present Knowledge and Trends*, Northern Illinois University.

Sponsel, Leslie E., Headland, Thomas N. and Bailey, Robert C. (eds) (1996), *Tropical Deforestation. The Human Dimension*, Columbia University Press, New York.

State Salinity Council (2000), *Natural Resource Management in Western Australia. The Salinity Strategy*, Government of Western Australia, Perth.

Tietenberg, T. (1988), *Environmental and Natural Resource Economics*, Scott, Foreman and Company, Glenview, Illinois.

Tisdell, C.A. (1991), *Economics of Environmental Conservation Economics for Environmental and Ecological Management*, Elsevier, Amsterdam, 233pp.

Todaro, Michael, P. (1977), *Economic Development in the Third World*, Longman, London and New York.

Tubiano, Celestina (1987), 'Synthetic Profile of Barangay Tabing, Tabango, Leyte', unpublished field report, ViSCA, Baybay, Leyte.

Tubiano, Celestina (1988), 'The Socio-Economic Constraints on the Adoption of Improved Cropping Methods by Upland Farmers in Leyte, Philippines. Site Report: Tabing, Tabango, Leyte (Site 4)', paper presented at the ACIAR project seminar, ViSCA, Leyte, September.

Tubiano, Celestina (1988), 'The Socio-Economic Constraints on the Adoption of Improved Cropping Methods by Upland Farmers in Leyte, Philippines, Site Report: Tabing, Tabango, Leyte (Site 4)', paper presented at the project conference, Visayas State College of Agriculture, Baybay, Leyte, September.

Tubiano, Celestina, Binoya, S., Reoma, B. and Tutor, N. (1988), 'Summary Notes of the Field Surveys by the Research Assistants', ViSCA, Baybay, Leyte.

Turner, R. Kerry (1991), 'Environmentally Sensitive Aid', in Pearce, David (ed), *Blueprint 2. Greening the World Economy*, Earthscan Publications, London.

Turner, R. Kerry (ed) (1990), *Sustainable Environmental Management. Principles and Practice*, Belhaven Press, London and Westview Press, Boulder, Colorado.

Tutor, Nilda (1987), 'Synthetic Profile of Barangay Sulpa (Canquiason) Villaba, Leyte', unpublished field report, ViSCA, Baybay, Leyte.

Tutor, Nilda (1988), 'The Socio-Economic Constraints on the Adoption of Improved Cropping Methods by Upland Farmers in Leyte, Philippines', paper presented at the project conference, Visayas State College of Agriculture, Baybay, Leyte, September.

Uitamo, E. (1996), 'Land Use History of the Philippines', in Palo, Matti and Mery, Gerardo (eds), *Sustainable Forestry Challenges for Developing Countries*, Environmental Science and Technology Library, Kluwer Academic Publishers, Dordrecht, Boston and London.

Ullsten, Ola and Rapport, David (2001), 'On the Politics of the Environment: Ecosystem Health and the Political Process', *Ecosystem Health*, Vol. 7(1), pp. 2-6.

Umali, Ricardo M. (1991) (Undersecretary of the Philippines Department of Environment and Natural Resources) in Blair, G. and Lefroy, R. (eds), *Technologies for Sustainable Agriculture on Marginal Uplands in Southeast Asia*, ACIAR Proceedings series no. 33.

Uy Eviota, Elizabeth (1992), *The Political Economy of Gender. Women and the Sexual Division of Labour in the Philippines*, Zed Books Ltd, London and New Jersey.

Van den Bergh, J.C.J.M and Nijkamp, D.P. (1991), 'Operationalizing Sustainable Development: Dynamic Ecological Economic Models', *Ecological Economics*, Vol. 4 (1).

Van Pelt, M.J.F. (1993), 'Ecologically Sustainable Development and Project Appraisal in Developing Countries', *Ecological Economics*, Vol. 7 (1).

Varian, Hal R. (1993), *Intermediate Microeconomics*, W.W. Norton, New York and London.

Velasco, Abraham, B. (1985), 'Traditional Farmer Practices Affecting Soil Erosion and Constraints to the Adoption of Soil Conservation Technology', in E.T. Craswell, J.V. Remenyi and L.G. Nallana (eds), *Soil erosion management*, ACIAR Proceedings series, no. 6.

Walker, Thomas S. and Ryan, James G. (1988), 'Against the Odds: Village and Household Economies in India's Semi Arid Tropics', ACIAR and ICRISAT, May.

Watson, Harold R. (1983), 'Developing a Hillside Farming Technique for Mindinao Hill Farmers', in *Proceedings, Hilly Land Development Workshop*, Ecotech Center, Cebu City, March.

Wernstedt, F.L. and Spencer, J.E. (1967), *The Philippine Island World. A Physical, Cultural and Regional Geography*, University of California Press, Berkeley and Los Angeles.

Worcester, Dean, C. (1921), *The Philippines Past and Present*, Macmillan, New York.

World Commission on Environment and Development (WCED) (1987), *Our Common Future*, Oxford University Press, Oxford.

Young, Michael, D. (1993), *For Our Childrens' Children. Some Practical Implications of Inter-Generational Equity, the Precautionary Principle, Maintenance of Natural Capital and the Discount Rate*, CSIRO Division of Wildlife and Ecology, Canberra.

ZDSPP (1983), *Developing Tropical Uplands*, Zamboanga del Sur Development Project, ACIL International Pty Ltd.

Appendix

Questions for the Perceived Discount Rate Survey

Interest rates proposed generally range from 110 % to 5 %.
Please circle the number next to the chosen response.

1] You have been offered a loan of p1000 to be paid back over 1 year. What is the most you would be prepared to pay as interest on the loan over the year?
1. 1100 pesos
2. 1000 pesos
3. 900 pesos
4. 800 pesos
5. 700 pesos
6. 600 pesos
7. 500 pesos
8. 400 pesos
9. 300 pesos
10. 200 pesos
11. 100 pesos
12. 50 pesos
13. More than any of the above (how much?)
14. Less than any of the above (how much?)

2] You have been offered a loan of p1000 for 5 months. What is the most you would be prepared to pay as interest on the loan over the 5 months?
1. 458 pesos
2. 417 pesos
3. 375 pesos
4. 333 pesos
5. 292 pesos
6. 250 pesos
7. 208 pesos

8. 167 pesos
9. 125 pesos
10. 83 pesos
11. 42 pesos
12. 21 pesos
13. More than any of the above (how much?)
14. Less than any of the above (how much?)

3] You have been offered a loan of p1000 for a period of 1 month. What is the most you would be prepared to pay as interest on the loan over the month?
1. 92 pesos
2. 83 pesos
3. 75 pesos
4. 67 pesos
5. 58 pesos
6. 50 pesos
7. 42 pesos
8. 33 pesos
9. 25 pesos
10. 17 pesos
11. 8 pesos
12. 4 pesos
13. More than any of the above (how much?)
14. Less than any of the above (how much?)

4] You have been offered a loan of 4 sacks of rice/corn for seed to be paid back over a year. How much would you be prepared to pay in interest for the loan (on top of repayments for the original amount) over the year?
1. 240 gantas
2. 200 gantas
3. 110 gantas
4. 100 gantas
5. 90 gantas
6. 80 gantas
7. 70 gantas
8. 60 gantas
9. 50 gantas
10. 40 gantas
11. 30 gantas
12. 20 gantas

13. More than any of the above (how much)?
14. Less than any of the above (how much)?

5] Would you rather:
>   1. Lend me 200 pesos for a year if I paid you back 220 pesos at the end of the year? Or,
>   2. Borrow from me 200 pesos for a year if you paid me back 220 pesos at the end of the year?

6] You have been offered a loan of p1000 to be paid back over 5 years. What is the most you would be prepared to pay each year as interest on the loan (on top of your repayments of the capital borrowed)?
1. 1100 pesos
2. 1000 pesos
3. 900 pesos
4. 800 pesos
5. 700 pesos
6. 600 pesos
7. 500 pesos
8. 400 pesos
9. 300 pesos
10. 200 pesos
11. 100 pesos
12. 50 pesos
13. More than any of the above (how much?)
14. Less than any of the above (how much?)

7] You have been offered a loan of 4 sacks of rice/corn for food to be paid back over 5 years. What is the most you would be prepared to pay each year as interest on the loan?
1.   110 gantas
2.   100 gantas
3.   90 gantas
4.   80 gantas
5.   70 gantas
6.   60 gantas
7.   50 gantas
8.   40 gantas
9.   30 gantas
10. 20 gantas
11. 10 gantas

12. 5 gantas
13. More than any of the above (how much?)
14. Less than any of the above (how much?)

8] You have been offered a loan of 4 sacks of rice to be paid back over 5 months. What is the most you would be prepared to pay as interest on the loan over the 5 months?
1. 46 gantas
2. 42 gantas
3. 37 gantas
4. 33 gantas
5. 29 gantas
6. 25 gantas
7. 21 gantas
8. 17 gantas
9. 12 gantas
10. 8 gantas
11. 4 gantas
12. 2 gantas
13. More than any of the above (how much?)
14. Less than any of the above (how much?)

9] You have been offered a loan of 200 pesos to be paid back over a period of 5 years. What is the most you would be prepared to pay in interest each year (on top of repayments)?
1. 220 pesos
2. 200 pesos
3. 180 pesos
4. 160 pesos
5. 140 pesos
6. 120 pesos
7. 100 pesos
8. 80 pesos
9. 60 pesos
10. 40 pesos
11. 20 pesos
12. 10 pesos
13. More than any of the above (how much?)
14. Less than any of the above (how much?)

10] You have been offered a loan of 200 pesos to be paid back over a period of 1 year. What is the most you would be prepared to pay in interest over the year (on top of repayments)?
1.   220 pesos
2.   200 pesos
3.   180 pesos
4.   160 pesos
5.   140 pesos
6.   120 pesos
7.   100 pesos
8.   80 pesos
9.   60 pesos
10. 40 pesos
11. 20 pesos
12. 10 pesos
13. More than any of the above (how much?)
14. Less than any of the above (how much?)

11] You have been offered a loan of 4 sacks of rice/corn for food to be paid back over 1 year. What is the most you would pay as interest on the loan for the year?
1.   110 gantas
2.   100 gantas
3.   90 gantas
4.   80 gantas
5.   70 gantas
6.   60 gantas
7.   50 gantas
8.   40 gantas
9.   30 gantas
10. 20 gantas
11. 10 gantas
12. 5 gantas
13. More than any of the above (how much?)
14. Less than any of the above (how much?)

12] You have been offered a loan of 200 pesos to be paid back over a period of 5 months. What is the most you would be prepared to pay in interest over the 5 months (on top of repayments)?
1.   92 pesos
2.   85 pesos

3. 75 pesos
4. 65 pesos
5. 58 pesos
6. 50 pesos
7. 42 pesos
8. 33 pesos
9. 25 pesos
10. 17 pesos
11. 8 pesos
12. 4 pesos
13. More than any of the above (how much?)
14. Less than any of the above (how much?)

13] Would you rather:

      1. Lend me 200 pesos for a year if I paid you back 300 pesos at the end of the year? Or,

      2. Borrow from me 200 pesos for a year if you paid me back 300 pesos at the end of the year?

14] You have been offered a gift of 1000 pesos. You can have the money now, or an increased amount in one year. If you were to wait for one year what would be the least you would accept in lieu of 1000 pesos now?

1. 1050 pesos
2. 1100 pesos
3. 1150 pesos
4. 1200 pesos
5. 1250 pesos
6. 1300 pesos
7. 1350 pesos
8. 1400 pesos
9. 1450 pesos
10. 1500 pesos
11. 1550 pesos
12. 1600 pesos
13. 1650 pesos
14. 1700 pesos
15. 1750 pesos
16. 1800 pesos
17. 1850 pesos
18. 1900 pesos
19. 1950 pesos

20. 2000 pesos
21. More than any of the above (how much?)
22. Less than any of the above (how much?)

15] You have been offered a loan of 20 gantas of rice/corn for food over 5 years. What is the most interest you are prepared to pay each year (on top of repayments)?
1.   22 gantas
2.   20 gantas
3.   18 gantas
4.   16 gantas
5.   14 gantas
6.   12 gantas
7.   10 gantas
8.   8 gantas
9.   6 gantas
10. 4 gantas
11. 2 gantas
12. 1 ganta
13. More than any of the above (how much?)
14. Less than any of the above (how much?)

16] You have been offered a loan of 20 gantas of rice/corn for food over 1 year. What is the most interest you are prepared to pay over that year (on top of repayments)?
1.   22 gantas
2.   20 gantas
3.   18 gantas
4.   16 gantas
5.   14 gantas
6.   12 gantas
7.   10 gantas
8.   8 gantas
9.   6 gantas
10. 4 gantas
11. 2 gantas
12. 1 ganta
13. More than any of the above (how much?)
14. Less than any of the above (how much?)

17] You have been offered a loan of 20 gantas of rice/corn for food over 5 months. What is the most interest you are prepared to pay over the 5 months (on top of repayments)?

1. 9.2 gantas
2. 8.3 gantas
3. 7.5 gantas
4. 6.5 gantas
5. 5.8 gantas
6. 5.0 gantas
7. 4.2 gantas
8. 3.3 gantas
9. 2.5 gantas
10. 1.7 gantas
11. 0.8 gantas
12. 0.4 gantas
13. More than any of the above (how much?)
14. Less than any of the above (how much?)

18] You have been offered a loan of 20 gantas of rice/corn for food over 1 month. What is the most interest you are prepared to pay over the month (on top of repayments)?

1. 1.8 gantas
2. 1.7 gantas
3. 1.5 gantas
4. 1.3 gantas
5. 1.2 gantas
6. 1.0 gantas
7. 0.8 gantas
8. 0.7 gantas
9. 0.5 gantas
10. 0.3 gantas
11. 0.2 gantas
12. 0.1 gantas
13. More than any of the above (how much?)
14. Less than any of the above (how much?)

19] Would you rather:

1. Lend me 1000 pesos for 3 months if I paid you back 1125 pesos at the end of 3 months? Or,

2. Borrow from me 1000 pesos for 3 months if you paid me back 1125 pesos at the end of 3 months?

20] You have been offered a loan of 200 pesos to be paid back over a period of 1 month. What is the most you would be prepared to pay in interest over the month (on top of repayments)?
1.   18.5 pesos
2.   16.5 pesos
3.   15 pesos
4.   13.5 pesos
5.   11.5 pesos
6.   10 pesos
7.   8.5 pesos
8.   6.5 pesos
9.   5.0 pesos
10. 3.5 pesos
11. 1.5 pesos
12. 0.85 pesos
13. More than any of the above (how much?)
14. Less than any of the above (how much?)

21] Suppose you are in a situation where if you don't borrow 1000 pesos now you will suffer severe financial in the near future. What interest would you be prepared to pay to borrow the money for a year?
1.   1100 pesos
2.   1000 pesos
3.   900 pesos
4.   800 pesos
5.   700 pesos
6.   600 pesos
7.   500 pesos
8.   400 pesos
9.   300 pesos
10. 200 pesos
11. 100 pesos
12. 50 pesos
13. More than any of the above (how much?)
14. Less than any of the above (how much?)

22] Suppose that you stand to make both a gain and a loss of equal value on your farm. You can either borrow to avert the financial loss on one commodity and in so doing forego the gain on another. Or you can borrow to make the gain on one commodity, but in so doing you incur a loss on the other. Which would you rather:

1. Borrow and repay 1000 pesos for a cost of 1000 pesos interest over a year to prevent a serious financial loss?
2. Borrow and repay 1000 pesos for a cost of 1000 pesos interest over a year to attract a good financial gain?
3. Borrow and repay 4 sacks of rice/corn for a cost of 4 sacks of rice/corn interest over a year to prevent a serious financial loss?
4. Borrow and repay 4 sacks of rice/corn for a cost of 4 sacks of rice/corn interest over a year to attract a good financial gain?

23] If you borrow 1000 pesos now you will make a very large profit on your next harvest. What amount of interest would you be prepared to pay to borrow the money for a year?
1.   1100 pesos
2.   1000 pesos
3.   900 pesos
4.   800 pesos
5.   700 pesos
6.   600 pesos
7.   500 pesos
8.   400 pesos
9.   300 pesos
10. 200 pesos
11. 100 pesos
12. 50 pesos
13. More than any of the above (how much?)
14. Less than any of the above (how much?)

24] If you borrow 1000 pesos now you can increase the number of livestock on your farm very profitably. How much interest would you be prepared to pay to borrow the money over the year?
1.   1100 pesos
2.   1000 pesos
3.   900 pesos
4.   800 pesos
5.   700 pesos
6.   600 pesos
7.   500 pesos
8.   400 pesos
9.   300 pesos
10. 200 pesos
11. 100 pesos

12. 50 pesos
13. More than any of the above (how much?)
14. Less than any of the above (how much?)

25] If you borrow 1000 pesos now you will make a very large profit on all your farm produce. How much interest are you prepared to pay on the loan over a year?
1.   1100 pesos
2.   1000 pesos
3.   900 pesos
4.   800 pesos
5.   700 pesos
6.   600 pesos
7.   500 pesos
8.   400 pesos
9.   300 pesos
10. 200 pesos
11. 100 pesos
12. 50 pesos
13. More than any of the above (how much?)
14. Less than any of the above (how much?)

26] Suppose you have money problems and if you don't borrow 1000 pesos now you won't be able to meet your mortgage payments. Someone has offered to lend you the money. What is the highest amount you would be prepared to pay to borrow the money for a year?
1.   1100 pesos
2.   1000 pesos
3.   900 pesos
4.   800 pesos
5.   700 pesos
6.   600 pesos
7.   500 pesos
8.   400 pesos
9.   300 pesos
10. 200 pesos
11. 100 pesos
12. 50 pesos
13. More than any of the above (how much?)
14. Less than any of the above (how much?)
27] Would you rather:

1. Borrow and repay 1000 pesos at a cost of 1000 pesos interest over 1 year?
2. Borrow and repay 1000 pesos at a cost of 500 pesos interest over 6 months?
3. Borrow and repay 1000 pesos at a cost of 250 pesos interest over 3 months?
4. Borrow and repay 1000 pesos at a cost of 83 pesos interest over 1 month?
5. Borrow and repay 4 sacks of rice/corn for food at a cost of 4 sacks rice/corn interest over 1 year?
6. Borrow and repay 4 sacks of rice/corn for food at a cost of 2 sacks of rice/corn interest over 6 months?
7. Borrow and repay 4 sacks of rice/corn for food at a cost of 25 gantas of rice/corn interest over 3 months?
8. Borrow and repay 4 sacks of rice/corn for food at a cost of 8.5 gantas of rice/corn interest over 1 month?

28] Suppose that you stand to make both a gain and a loss of equal value on your farm. You can either borrow to avert the financial loss on one commodity and in so doing forego the gain on another. Or you can borrow to make the gain on one commodity, but in so doing you incur a loss on the other. Which would you rather:

1. Borrow and repay 1000 pesos for a cost of 500 pesos interest over a year to prevent a serious financial loss on the farm?
2. Borrow and repay 1000 pesos for a cost of 500 pesos interest over a year to attract a good financial gain on the farm?
3. Borrow and repay 4 sacks of rice/corn for food for a cost of 2 sacks of rice/corn over a year to prevent a serious financial loss on the farm?
4. Borrow and repay 4 sacks of rice/corn for food for a cost of 2 sacks of rice/corn over a year to attract a good financial gain on the farm?

29] Suppose that you stand to make both a gain and a loss of equal value on your farm. You can either borrow to avert the financial loss on one commodity and in so doing forego the gain on another. Or you can borrow to make the gain on one commodity, but in so doing you incur a loss on the other. Which would you rather:

1. Borrow and repay 1000 pesos for a cost of 100 pesos over a year to prevent a serious financial loss on the farm?

2. Borrow and repay 1000 pesos for a cost of 100 pesos over a year to attract a good financial gain on the farm?

3. Borrow and repay 4 sacks of rice/corn for food for a cost of 10 gantas of rice/corn to prevent a serious financial loss on the farm?

4. Borrow and repay 4 sacks of rice/corn for food for a cost of 10 gantas of rice/corn over a year to attract a good financial gain on the farm?

30] You have been offered a gift of 200 pesos. You can have the money now, or an increased amount in one year. If you were to wait for one year what would be the least you would accept in lieu of 200 pesos now?

1.   210 pesos
2.   220 pesos
3.   230 pesos
4.   240 pesos
5.   250 pesos
6.   260 pesos
7.   270 pesos
8.   280 pesos
9.   290 pesos
10. 300 pesos
11. 310 pesos
12. 320 pesos
13. 330 pesos
14. 340 pesos
15. 350 pesos
16. 360 pesos
17. 370 pesos
18. 380 pesos
19. 390 pesos
20. 400 pesos
21. More than any of the above (how much?)
22. Less than any of the above (how much?)

31] Would you rather:

1. Lend me 1000 pesos for a year if I paid you back 1500 pesos at the end of the year? Or,

2. Borrow from me 1000 pesos for a year if you paid me back 1500 pesos at the end of the year?

32] You have been offered a loan of 4 sacks of rice/corn for food over a period of 1 month. What is the most you would be prepared to pay in interest for the loan over the month?
1. 9 gantas
2. 8 gantas
3. 7.5 gantas
4. 6.6 gantas
5. 5.8 gantas
6. 5 gantas
7. 4 gantas
8. 3.3 gantas
9. 2.5 gantas
10. 1.7 gantas
11. 0.8 gantas
12. 0.4 gantas
13. More than any of the above (how much?)
14. Less than any of the above (how much?)

33] Would you rather:
> 1. Lend me 1000 pesos for a year if I paid you back 1100 pesos at the end of the year? Or,
> 2. Borrow from me 1000 pesos for a year if you paid me back 1100 pesos at the end of the year?

34] Would you rather:
> 1. Lend me 1000 pesos for 3 months if I paid you back 1025 pesos at the end of 3 months? Or,
> 2. Borrow from me 1000 pesos for 3 months if you paid me back 1025 pesos at the end of 3 months?

35] Would you rather:
> 1. Lend me 200 pesos for 3 months if I paid you back 205 pesos at the end of 3 months? Or,
> 2. Borrow from me 200 pesos for a year if you paid me back 205 pesos at the end of 3 months?

36] You have very serious financial problems and need money straight away. You cannot borrow elsewhere so I offer to lend you 1000 pesos for a year, but only if you will give me a total of 2000 pesos at the end of the year. Would you:
> 1. Accept my offer?

2. Reject my offer?

37] You have very serious financial problems and need money straight away. You cannot borrow elsewhere so I offer to lend you 1000 pesos for 5 months, but only if you will give me a total of 2083 pesos at the end of 5 months. Would you:
     1. Accept my offer?
     2. Reject my offer?

38] You have very serious financial problems and need money straight away. You cannot borrow elsewhere so I offer to lend you 1000 pesos for a year, but only if you will give me a total of 3500 pesos at the end of the year. Would you:
     1. Accept my offer?
     2. Reject my offer?

39] You have very serious financial problems and need money straight away. You cannot borrow elsewhere so I offer to lend you 1000 pesos for 5 months, but only if you will give me a total of 2708 pesos at the end of 5 months. Would you:
     1. Accept my offer?
     2. Reject my offer?

40] Would you rather:
     1. Lend me 200 pesos for 3 months if I paid you back 225 pesos at the end of 3 months? Or,
     2. Borrow from me 200 pesos for a year if you paid me back 225 pesos at the end of 3 months?

41] You have been offered a loan of 10,000 (ten thousand) pesos to be paid back over a period of a year. What is the most you would be prepared to pay in interest on the loan (on top of the original amount borrowed) over the year?
1.   11,000 pesos
2.   10,000 pesos
3.   9,000 pesos
4.   8,000 pesos
5.   7,000 pesos
6.   6,000 pesos
7.   5,000 pesos
8.   4,000 pesos

9. 3,000 pesos
10. 2,000 pesos
11. 1,000 pesos
12. 500 pesos
13. More than any of the above (how much)?
14. Less than any of the above (how much)?

Notes:
1 *cavan* (sack) of rice (*palay*) is roughly equivalent to 37.5 kg
4 sacks of rice (palay) is roughly equivalent to 100 gantas
1 ganta of rice = 1.5 kg
4 sacks of rice (palay) is worth about 1000 pesos
Palay is unmilled rice

Index

Printed and bound by CPI Group (UK) Ltd, Croydon, CR0 4YY

22/10/2024

01777625-0010